Skin Penetration: Hazardous Chemicals at Work

by
Philippe Grandjean, M.D.
Odense University, Denmark

A report prepared for
the Commission of the European Communities, Directorate-General
Employment, Industrial Relations and Social Affairs, Health and
Safety Directorate

Taylor & Francis
London • New York • Philadelphia
1990

| UK | Taylor & Francis Ltd, 4 John St., London WC1N 2ET |
| USA | Taylor & Francis Inc., 1900 Frost Rd., Suite 101, Bristol, PA 19007 |

© ECSC-EEC-EAEC, Brussels–Luxembourg, 1990

LEGAL NOTICE
Neither the Commission of the European Communities nor any person acting on behalf of the Commission is responsible for the use which might be made of the following information.

All rights reserved. No part of this publication may be reproduced, stored in a retrieval system, or transmitted, in any form or by any means, electronic, electrostatic, magnetic tape, mechanical, photocopying, recording or otherwise, without the prior permission of the copyright owner and publishers.

British Library Cataloguing in Publication Data
Grandjean, Philippe
  Skin penetration: hazardous chemicals at work.
  1. Dangerous industrial chemicals. Safety aspects
  I. Title
  363.179

  ISBN 0-85066-834-4

Library of Congress Cataloguing-in-Publication Data is available

Cover design by Jordan and Jordan, Fareham, Hampshire

Typeset in 10/12 English Times by
Chapterhouse, Formby

Printed in Great Britain by Burgess Science Press, Basingstoke on paper which has a specified pH value on final paper manufacture of not less than 7.5 and is therefore 'acid free'.

Skin Penetration:
Hazardous Chemicals at Work

Publication No. EUR 12599 EN of the Commission of the European Communities, Scientific and Technical Communication Unit, Directorate-General Telecommunications, Information Industries and Innovation, Luxembourg

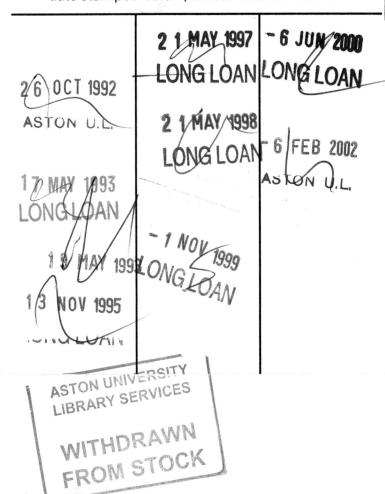

Contents

	Preface	ix
Section A	**Percutaneous absorption**	
Chapter 1.	**Percutaneous Absorption**	3
	The skin as a barrier	3
	Normal functions of the skin	3
	Variations in structure	4
	Skin exposure at work	6
	Past experiences	6
	Current significance	7
	Test methods	8
	Application of Fick's law	8
	Physicochemical properties	11
	Experimental data	12
	Human data	14
	Potentials for classification	15
	Chemical properties	15
	Partition coefficients	17
	Alternative pathways	17
	Structure–activity relationships	18
	Sources of variation	18
	The vehicle	18
	Hydration of the skin	18
	Integrity of the skin	19
	Damaging effects of chemicals	20
	Skin denotation	20
	Reasons for a 'skin' denotation	20
	Current practice	21
	International variations	22
	Recent modifications	23

	Selection of chemicals for review	24
	Validity of 'skin' denotation	24
	Structure for evaluation	26
	Preventive approaches	27
	Planning for prevention	27
	Treatment	27
	Protective clothing	28
	Barrier creams	29
	References	29

Section B Chemicals as percutaneous hazards

Chapter 2.	**Inorganic and organometal compounds**	37
	Introduction	37
	Hydrogen cyanide and cyanide salts	39
	Tetraethyllead	40
	Thallium	42
	Other inorganic compounds	43
	References	44
Chapter 3.	**Simple aliphatic compounds**	47
	Introduction	47
	Aliphatic hydrocarbons	49
	Methanol	49
	Allyl alcohol	51
	Other alcohols	52
	Ketones	53
	Ethylene glycol, monomethyl ether	54
	Other glycol derivatives	55
	References	56
Chapter 4.	**Halogenated aliphatic compounds**	59
	Introduction	59
	Carbon tetrachloride	60
	Methyl bromide	61
	Methyl iodide	63
	Other compounds	64
	References	66
Chapter 5.	**Aliphatic amides, nitriles and amines**	69
	Introduction	69
	N,N-Dimethylformamide	70
	N,N-Dimethylacetamide	71
	Acrylamide	73

	Acrylonitrile	75
	Other compounds	76
	References	77
Chapter 6.	**Isocyclic hydrocarbons, alcohols and related compounds**	81
	Introduction	81
	Benzene	82
	Toluene	84
	Xylenes	86
	Phenol	88
	Cresols	89
	Other compounds	91
	References	92
Chapter 7.	**Halogenated cyclic compounds**	97
	Introduction	97
	Polychlorinated biphenyls	98
	Chlorinated naphthalenes	100
	Pentachlorophenol	101
	Toxaphene	103
	Other chlorinated aromatic compounds	104
	Other chlorinated cyclic compounds	106
	References	108
Chapter 8.	**Isocyclic amines**	113
	Introduction	113
	Aniline	114
	4-Nitroaniline	116
	Other compounds	117
	References	118
Chapter 9.	**Organic nitro compounds and nitrates**	121
	Introduction	121
	Nitroglycol	122
	Nitroglycerin	123
	Tetryl	124
	Nitrobenzene	125
	2,4,6-Trinitrotoluene	127
	Picric acid	128
	4,6-Dinitro-*o*-cresol	129
	Other compounds	130
	References	131

Chapter 10.	**Other nitrogen compounds**	135
	Introduction	135
	Carbaryl	136
	Nicotine	138
	Paraquat	139
	Hydrazine	141
	1,1-Dimethylhydrazine	142
	Phenylhydrazine	143
	Other compounds	145
	References	146
Chapter 11.	**Heterocyclic oxygen and sulfur compounds**	151
	Introduction	151
	Epichlorohydrin	152
	1,4-Dioxane	153
	Other compounds	154
	References	155
Chapter 12.	**Organophosphorus compounds**	157
	Introduction	157
	Dichlorvos	158
	Malathion	159
	EPN	161
	Parathion	162
	Diazinon	164
	Other compounds	165
	References	167
Chapter 13.	**Organic sulfur compounds**	171
	Introduction	171
	Carbon disulfide	171
	Dimethyl sulfate	173
	Dimethyl sulfoxide	174
	Other compounds	176
	References	176
Chapter 14.	**Conclusions**	179
	Index	181

Preface

This book is based on a project funded by the Commission of the European Communities (CEC). A large part of the scientific literature retrieval and the compilation on 'skin' notations was carried out by the International Register of Potentially Toxic Chemicals (IRPTC). Other useful data, e.g. on chemical production, was retrieved through the Environmental Chemicals Data Information Network (ECDIN). The author was responsible for preparing all reports generated under this project through the IRPTC during the period 1983 to 1989. The present volume is revised and updated from the reports submitted to the CEC.

In occupational health, contamination of the skin with chemicals has received comparatively little interest in the past. The reasons for the lack of attention to this problem may include the fact that uptake through the skin is difficult to measure, and that the possible toxic significance is hard to evaluate. However, there is increasing evidence that absorption of toxic chemicals through the skin is of relevance. In particular, with the improved protection against inhalation of such chemicals, the percutaneous uptake will tend to become more important. A need was therefore detected for a thorough review of the evidence available for percutaneous uptake and possible suggestions for safety practices in the workplace.

The book contains a general section which outlines the mechanism of percutaneous absorption and the methods used to evaluate its significance. The rest of the book reviews the different classes of chemicals with emphasis on those compounds that are considered major skin hazards. Hopefully, this book will inspire increased efforts to prevent percutaneous absorption.

Section A
Percutaneous absorption

Chapter 1
Percutaneous absorption

The skin as a barrier

Normal functions of the skin

Healthy skin is a remarkably good barrier which limits loss of water and other essential compounds from the body while diminishing the penetration of potentially toxic substances into the body (Scheuplein and Bronaugh, 1983). However, the barrier function is not permanent, nor is it constant at all times. Penetration of the skin is a passive process, that occurs unaided by the cells within the skin. With different chemical substances, the penetration rate may vary by as much as four orders of magnitude. In addition, damage or disease may render the barrier discontinuous, with a resulting increase in percutaneous absorption of several orders of magnitude.

The skin covers about 1.8 m^2 of body surface and consists of three main layers, the epidermis, the dermis and the subcutaneous fat (figure 1.1). The outer, horny layer (stratum corneum) consists of several layers of flattened, keratinized, dead cells. The deeper layers of the epidermis consist of cell layers that exhibit increasing degrees of flattening and keratinization toward the stratum corneum. Below the stratum corneum, the dermis (or corium) forms papillae that contain blood vessels. Percutaneous absorption would therefore entail a transfer of the chemical from the surface of the skin to the blood vessels in the dermis. The bulk of evidence shows that the stratum corneum constitutes the main barrier against percutaneous absorption (Scheuplein and Bronaugh, 1983).

Thus, percutaneous absorption involves a series of processes occurring in sequence. First, molecules of the chemical must be absorbed at the surface of the stratum corneum and then diffuse through the flattened cell layers; then the compound must enter the viable epidermis, then the papillary dermis until it reaches a capillary where it enters the systemic circulation.

Water penetrates the skin, e.g. as insensible perspiration. The penetration rate is similar for saturated water vapour and water applied to the skin and averages about 0.2–$0.4 \,\mathrm{mg\,cm^{-2}\,h^{-1}}$ at 30°C (Scheuplein and Bronaugh, 1983). Hydrophilic chemicals generally penetrate the skin more slowly than water. Cell membranes and intercellular spaces contain lipids, thus making penetration possible also for lipophilic compounds. The skin is a biological tissue with complex properties that distinguish this barrier from a simple membrane (Wester and Maibach, 1983).

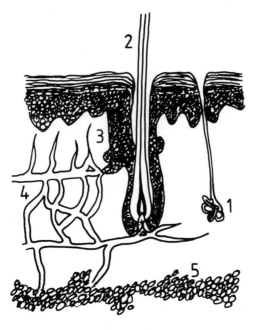

Fig. 1.1. Schematic structure of normal skin. The stratum corneum is the most superficial layer of the epidermis (dark cells) and consists of flattened cells. The skin appendages include sweat glands (1), hairs (2) and sebaceous glands (3). The dermis consists mainly of connective tissue and has a supply of blood vessels (4). The deeper subcutis contains lipid cells (5).

When chemical substances penetrate the skin, local toxicity may occur within the skin, as with caustic, allergenic and phototoxic agents or compounds that cause skin cancer. Further, the chemicals may be more toxic than if absorbed from an oral dose, as detoxification in the liver may be bypassed. On the other hand, metabolizing enzymes in the skin may also affect the systemic toxicity of chemicals absorbed by this route. The skin contains many of the same enzymes that also occur in the liver, and epidermal enzyme activities can be of the same order of magnitude as hepatic levels (Noonan and Wester, 1985). For example, a large percentage of dermally applied organophosphorous pesticides is metabolized during the passage through the skin (Frederiksson *et al.*, 1961; van Hooidonk *et al.*, 1980). On the other hand, aryl hydrocarbon hydroxylase in the skin may convert noncarcinogenic benzo[*a*]pyrene into a potent carcinogen that could cause effects within the skin or elsewhere in the body (Wester and Maibach, 1984). Thus, first-pass metabolism may therefore significantly decrease or increase the systemic bioavailability of the compound. For such reasons, dermal application is considered necessary in routine toxicity testing programmes.

Variations in structure

The skin also contains so-called appendages, e.g. sweat ducts and hair follicles (figure 1.1). They constitute a relatively minor fractional area of the skin, i.e.

about 10^{-2}–10^{-3}. Passage through the appendages has been shown to occur, but it only plays a role for substances that otherwise move very slowly through the stratum corneum (Scheuplein and Bronaugh, 1983). Thus, skin appendages may under some circumstances provide an important pathway.

The thickness of the horny layer varies considerably. The thinnest layer (eyelids or scrotum) is less than 0.01 mm, while the thickest (palms and soles) is 1 mm (Scheuplein and Bronaugh, 1983). Due to a different structure, however, the palmar and plantar horny pads are at least as permeable to water and several other substances as is the thinner stratum corneum covering the rest of the body (figure 1.2). Thus, both thickness and diffusivity play a role for the rate of percutaneous absorption. However, thickness is of particular importance in one aspect. The time for the initial diffusion front to pass through a membrane depends on the square of the membrane thickness (Scheuplein and Bronaugh, 1983).

The absorption rate through the skin of the ventral forearm is frequently used as a reference. However, penetration of, e.g. malathion, is faster through the horny pads, even faster when applied on the scalp or facial skin and fastest through scrotal epidermis; other pesticides may show somewhat different variations with the site of application (Maibach *et al.*, 1971). In general, penetration rates through various skin areas can be ranked as follows: plantar >

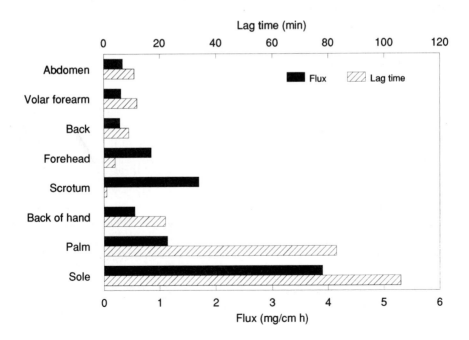

Fig. 1.2. Percutaneous absorption rate and lag time for water penetration through skin form different anatomical locations.
Data from Scheuplein and Bronaugh, 1983.

scrotal > palmar > dorsum of hand > forehead and scalp > arms, legs, trunk (figure 1.2) (Scheuplein and Bronaugh, 1983). Additional information on penetration of water and topical drugs through the skin suggests that regional differences may vary according to the chemical encountered.

The entire stratum corneum is replaced about every 2 weeks in adults (Scheuplein and Bronaugh, 1983). Thus, a continuous shedding of cells occurs. For substances with a slow penetration rate, but high solubility in the stratum corneum, an initial absorption into the outermost layers of the skin may then be followed by a release as the cells are shed.

In the workplace, some skin contact with chemicals is almost unavoidable. Such cutaneous exposures may not necessarily constitute a hazard, but seemingly minimal skin contact can sometimes lead to toxic effects. Many factors play an important role with regard to the rate and extent of absorption through the skin. Current research in this area is just beginning to define the influence of a variety of parameters on skin absorption. Because of the complexities, individual percutaneous uptake cannot be readily predicted, even from detailed information on working methods and habits. This situation is very different from the more familiar respiratory exposure to airborne chemicals. However, for preventive purposes, safe working practices and appropriate safeguards must be defined for each specific exposure situation. Such preventive guidelines require a detailed and prudent evaluation of the information on percutaneous absorption and its variations under different conditions.

Skin exposure at work

Past experiences

A main objective of occupational health practice is to secure safe production methods and working conditions. In assessing compliance, emphasis is placed on surveillance of air concentrations in relation to exposure limits. This reliance on airborne concentrations is due to the practicality of such measurements and the fact that respiratory intake of toxic substances has appeared to be the main problem.

However, many chemicals can be absorbed through the skin in toxic amounts, and such hazards may well occur when airborne concentrations of the chemical are within legal limits. Percutaneous absorption may be difficult to detect, because it is hard to measure, and the effects may not be readily apparent. Thus, the systemic significance of percutaneous absorption has probably been underestimated in industrial experience and in occupational health practice.

However, visible effects of skin absorption have been amply demonstrated by past experience. For example, uptake through the skin can cause local effects, such as allergy or discolouration. More seriously, extensive percutaneous absorption has occasionally caused serious poisoning cases. However, absorption through the skin is rather difficult to document, unless extensive skin exposure took place and other means of intake can be excluded or disregarded. A few examples will illustrate the types of evidence available.

With acrylamide, a water-soluble, solid compound, considerable skin exposure to the dusts in a polymer factory led to chronic polyneuropathy in several workers (Garland and Patterson, 1967). The fact that the compound did not occur as a liquid or in solution may erroneously have suggested that percutaneous absorption was unlikely. Also, that the effects occurred with a considerable delay may have suggested other causation than the acrylamide exposure.

In contrast, during dynamite production, dermal exposure to nitroglycol (ethylene glycol dinitrate) has in a matter of few minutes resulted in acute symptoms with a very obvious relation between skin contact and systemic toxicity (Hogstedt and Ståhl, 1980). In this case, the relation between exposure and effect was readily apparent. However, the reason for surprise was that even limited cutaneous exposure could cause severe effects with a minimal delay.

Perhaps the most serious skin exposure hazard has occurred in relation to pesticide application: drenched clothes, lack of protective equipment and unsafe spraying methods have resulted in considerable numbers of intoxications, mainly due to percutaneous absorption (World Health Organization (WHO), 1982). One study of pesticide applicators suggested that respiratory exposure averaged less than 1% of the dermal exposure (Wolfe *et al.*, 1967). Even in tractor drivers applying pesticides, respiratory exposures may be of negligible significance when compared to percutaneous absorption (Jegier, 1974). Dermal uptake may be a protracted process, and persistence of certain pesticides on the hands has been demonstrated (Kazen *et al.*, 1974). These observations are particularly important, as pesticides are primarily used in hot climates that may deter applicators from wearing protective clothing.

These few examples serve to illustrate that compounds from a variety of activities may pose a potential hazard due to dermal exposure, and that such exposure may occur under widely differing circumstances.

Current significance

Although the above examples reflect past exposure situations, working conditions have not uniformly improved such that similar cases could not occur again. Percutaneous absorption still plays an important role, perhaps even more so as respiratory exposure is being limited. In some cases, airborne concentrations may be of limited relevance. Thus, the documentation of skin absorption of two pesticides, carbaryl and malathion, was so extensive that a WHO expert group recommended major emphasis on skin protection and refrained from suggesting a health-based respiratory exposure limit (WHO, 1982). This decision was due to: (i) the overwhelming importance of skin absorption; and (ii) the lack of documentation on a respiratory exposure–response relationship where interference from percutaneous absorption could be excluded. A similar situation may apply to several other chemicals.

Percutaneous absorption is important in one additional situation. For an increasing number of substances, biological monitoring is being used. Thus, the concentration of the chemical or its metabolite in blood or urine is used as an indication of the exposure and resulting uptake. Often, a blood sample is taken from the cubital vein of the arm. For some solvents, exposure of the hand to the solvent may lead to erroneously high dosage estimations when judged from concentrations of the chemical in the blood draining the site of absorption (Aitio et al., 1984).

Percutaneous exposure can be determined by several methods, none of which is practical for routine use. A skin wipe can remove a certain amount of retained chemicals from the skin, possibly as an indication of recent exposures, particularly to lipophilic compounds with a slow penetration rate, such as polychlorinated biphenyls (PCB) (Lees et al., 1987), polycyclic aromatic hydrocarbons (Wolff et al., 1989) or certain pesticides (Kazen et al., 1974). However, the validity of the skin wipe technique remains to be determined. The WHO (1986) standard protocol for studies of pesticide exposures recommends disposable overalls and gauntlets or cellulose pads attached to the clothing or skin; after completion of the spraying, these materials are then analyzed for pesticide contents. The patch method has been most frequently used in studies of pesticide exposure, although the validity is not well documented. A more recent approach is to add a fluorescent tracer to a pesticide formulation, and dermal deposition is then determined by the UV-fluorescence detected by an image-processing system (Fenske, 1988). For experimental studies, this method appears to be promising, though it requires expensive equipment. Thus, methods are available to determine the extent of skin deposition in occupational situations or model experiments. However, they cannot be used as a routine method to assess whether or not preventive efforts are sufficient.

Test methods

Application of Fick's law

Much of the knowledge concerning percutaneous absorption is derived from studies of cosmetics and pharmacological agents used in topical preparations (Grasso and Landsdown, 1972; Nater and de Groot, 1983; Wester and Maibach, 1983, 1985). Such exposures are intentional and meant to be beneficial, but the basic mechanisms involved may to some degree apply to accidental dermal exposures to chemical substances in the workplace.

The theoretical basis of diffusion and transpaort through membranes is useful for the understanding of percutaneous absorption. Although Fick's law only applies to very specific circumstances, in many cases it will give a reasonable approximation. Thus, the steady-state penetration flux J can be described by the relation

$$J = k_p \, dC$$

where k_p is a proportionality constant (the permeability constant or coefficient) and dC the concentration difference across the membrane. Based on several assumptions that seem to apply in most instances,

$$k_p = \frac{K_{MV} D_M}{\delta_M}$$

where K_{MV} is the partition coefficient between stratum corneum (membrane) and water (vehicle), D_M the diffusion constant and δ_M the thickness of stratum corneum. Thus, the permeability constant increases with the solubility of the compound in stratum corneum, and it decreases with increased thickness of the skin. The flux through the stratum corneum then becomes

$$J = \frac{K_{MV} D_M \, dC}{\delta_M}$$

This expression can be applied to many experimental conditions, although it may not always be accurate. The passage of a chemical through the skin membrane is illustrated in figure 1.3.

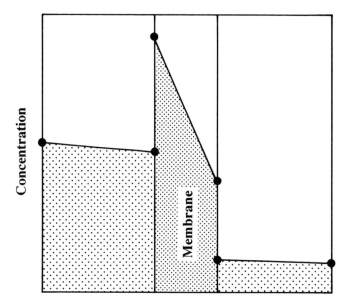

Fig. 1.3. Concentration gradients illustrating Fick's law. The concentration is higher on the outside (left) than inside (right). The concentration in the membrane is higher than in the medium as a result of a higher affinity. The steepness of the gradient will depend upon the penetration rate.

The permeability constant is usually given in $cm \, h^{-1}$, and the k_p for water is about $10 \times 10^4 \, cm \, h^{-1}$. Most chemicals dealt with in this volume have a lower permeability constant than water, thus indicating a slower penetration of the skin. However, with certain vehicles, the permeability constant can increase considerably.

The theory of membrane penetration has advanced through studies of simple, artificial membranes (Anderson *et al.*, 1972; Gary–Bobo *et al.*, 1969). However, in light of the complexity of human skin, the interpretation of such studies is difficult. Mathematical modelling of skin penetration is possible but may become rather complicated under non-steady-state conditions, in particular when more than one diffusion barrier is apparent, or when metabolic breakdown occurs in the skin (Guy and Hadgraft, 1985; Scheuplein, 1978).

A latency period will occur before diffusion of the chemical can be detected on the other side of the skin barrier. This latency is usually called the breakthrough time (figure 1.4).

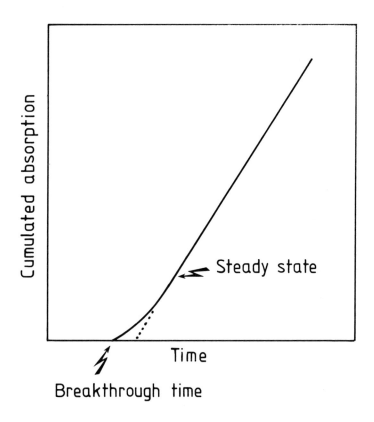

Fig. 1.4. Temporal pattern of skin penetration. Due to the time needed for penetration of the membrane (e.g. the skin), absorption is delayed; absorption cannot be detected before the breakthrough time. The dotted line is an extension of the straight line obtained at steady state; the lag time is defined by the intersection of the dotted line with the horizontal axis.

Physicochemical properties

As the flux through the skin depends on K_{MV}, solubility in the stratum corneum plays a major role in skin penetration. Solubilities have been expressed in several ways. First of all, the water–octanol partition coefficient (P_{ow}) is frequently available and is of some use; often this parameter is given as the $\log_{10}$ value (Hansch and Leo, 1979; Leo et al., 1971). However, the stratum corneum and octanol have somewhat different solubility characteristics. Thus, octanol has a permeability constant of 0.1×10^4 cm h^{-1}, but when applied in aqueous solution the k_p increases to 52×10^4 cm h^{-1}, i.e. in excess of the k_p for water (Scheuplein, 1978). With chlorinated organic solvents, a P_{ow} close to unity seems to be associated with a faster skin penetration rate (Tsuruta, 1975; 1982).

In general, a high k_p is seen with compounds that retain some hydrophilicity and thus have a P_{ow} in the medium range. However, as the chemical may not necessarily occur in aqueous solution, the use of P_{ow} becomes complicated. On the other hand, the P_{ow} may be a useful parameter in predicting toxicity and persistence in the body (Hansch et al., 1989), and the partition coefficient may therefore be of more general toxicological interest. Instead of octanol as a reference for solubility, several other organic compounds have been used.

Empirical studies have suggested that the skin penetration rate (in mg cm^{-2} h^{-1}) can be estimated by the following equation:

$$J = \frac{C_{sat}}{15}(0.038 + 0.153 \times P_{ow})\,e^{-0.016 \times MW}$$

where C_{sat} is the concentration of the compound (in mg ml^{-1}) in a saturated aqueous solution and MW the molecular weight of the compound (Fiserova-Bergerova and Pierce, 1989). In addition to the hydrophilic and lipophilic properties, this equation also takes the size of the diffusion unit into account. However, only a preliminary approximation can be expected.

More specifically, the solubility in stratum corneum can be assessed by using various types of material. Human callous tissue has been suggested for this purpose (Wurster, 1978; Wester et al., 1987). Further, extensive information is available on the affinities of organic solvents to psoriasis scales; many of these affinities have been determined by a swelling test, and others have been calculated from solubility parameters (Hansen and Andersen, 1988). Although callous tissue and psoriasis scales may differ from normal stratum corneum, such data provide at least semi-quantitative information on solubility in stratum corneum. For example, a high solubility has been predicted for aniline, dimethyl sulfoxide (DMSO), ethylene dibromide, nitrobenzene, and 1,1,2,2-tetrabromoethane (Hansen and Andersen, 1988).

Recently, a new approach to skin penetration has been suggested. Rougier et al. (1983) presented results that indicate a close correlation between percutaneous absorption and the uptake in stratum corneum within 30 min. This correlation is plausible, as the absorption rate depends on K_{MV} (see above). The experimental design included a 30-min application followed by immediate washing of the application site; the upper layers of stratum corneum were then removed by six successive strippings using adhesive tape. The concentration of a variety of substances in the stripped cells correlated closely to the total amount absorbed and reflected species differences (Dupuis *et al.*, 1984), the influence of application time (Rougier *et al.*, 1985) and vehicle (Dupuis *et al.*, 1986) as well as differences between regions of the human body (Rougier *et al.*, 1986). This technique could simplify tests for percutaneous absorption. However, binding or metabolism in the stratum corneum as well as possible effects of occlusion or evaporation may possibly influence the validity of test results. Although promising, this method has so far been used for very few compounds of limited interest in occupational health.

Experimental data

Likewise, *in vitro* studies using excised (viable or non-viable) skin have provided much useful information on penetration rates of a considerable number of compounds under various circumstances. A model for studying skin penetration *in vitro* is shown in figure 1.5. The critical problems include the choice of species, the age, the site from which the skin is obtained, the mounting of the skin under

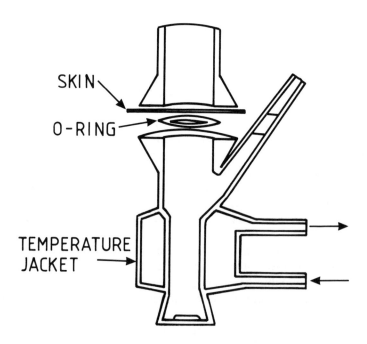

Fig. 1.5. Experimental cell for studying *in vitro* skin penetration. The skin is fastened with an O-ring, and the study solution is applied on the outer side (top). A steady temperature is maintained by a flow of water in the temperature jacket. The chamber below the skin membrane is designed so that samples can be obtained for analysis.

the experimental conditions, the method of applying the chemical on the surface of the skin, and the removal of penetrating amounts on the receptor side of the skin (Bronaugh and Maibach, 1985; Franz, 1975). These parameters have been dealt with in different ways in past studies, thus causing sufficient variability to limit the value for estimating penetration rates and health significance of percutaneous absorption. The major limitation in these studies is that the receptor side of the excised skin is exposed to an aqueous solution, rather than dermal tissues with a blood supply. Thus, extrapolation to dermal absorption in humans must be cautious.

Autoradiography has been used for limited purposes and has documented the penetration of certain chemicals into the skin and particular structures of the skin. Also, skin biopsy followed by chemical analysis has been used to determine penetration of a chemical into the skin. Such methods can only provide information on the distribution within the skin at a particular point of time. Thus, quantitative penetration rates cannot be displayed by these approaches.

A new experimental approach has been developed that solves the problem of accessible blood supply beneath the skin. This model uses the nude rat and the abdominal skin that has an independent blood supply from the superficial epigastric vessels. A skin sandwich is generated as a flap by attaching a split-thickness skin graft to the subcutaneous surface of the skin; the flap is then isolated with the vasculature, transferred to the back of the animal and sutured in place (Wojciechowski *et al.*, 1987). This island skin flap can then be exposed by dermal application, and the absorption can be determined from analysis of blood from the vessels at the root of the flap.

Pig skin has been used for another experimental model. In this case, a tubed skin flap is formed from the gastric skin; at a second surgical procedure, this tube with its vascular supply is removed and transferred to a perfusion apparatus (Riviere *et al.*, 1986). A 5 cm^2 area is available for dermal exposure, and the flux can be determined from analyses of the perfusate. Early experience with organophosphates and other compounds indicates that this model may have a high degree of validity (Carver *et al.*, 1989). However, the perfusate may not be sufficiently lipophilic to mimic *in vivo* conditions. With both of these models, microsurgical skills are needed, but the information gathered may well render the investment worthwhile.

In experiments using intact animals, dermal absorption has been measured by several techniques. By the disappearance technique, or analysis by difference, a known amount of the compound is applied to the skin, and the disappearance from the application site is followed (Grasso and Lansdown, 1972). The use of radioactively labelled compounds has greatly simplified the techniques. However, evaporation is a possible source of error; if prevented by occlusion, the occlusion may then significantly affect skin penetration.

The absorption may also be determined by analysis of concentrations in the blood and excreta. In this case, dilution may affect the results if the chemical occurs endogenously in the body.

The above methods are useful in quantifying the degree or rate of percutaneous absorption. However, some compounds are rapidly excreted after absorption, and some have a low degree of toxicity. Thus, a fast skin penetration may not necessarily mean that skin contact can cause systemic toxicity. In contrast, some compounds that only pass through the skin very slowly may accumulate in the body and result in chronic adverse effects. Also, some of these chemicals may be carcinogenic, so that even a minimal uptake through the skin would constitute a hazard.

A range of effect parameters have been used to quantify the significance of skin penetration. Specific effects include, e.g. cholinesterase inhibition due to organophosphorus pesticides. In more common use, the dermal LD_{50} refers to the amount of the chemical that induces a 50% mortality by a particular method of application to the skin of a species for a given length of time. The circumstances of past studies in this area rarely comply with modern requirements for good laboratory practice. For example, the use of occlusion is sometimes not indicated. Slight differences with regard to this factor or to concentration, surface area treated, vehicle, method of removal after cessation of exposure, and other parameters may significantly influence the rate of absorption. Obviously, a relatively low dermal LD_{50}, in particular when compared with LD_{50} values obtained by oral or intraperitoneal exposure, would tend to indicate that skin absorption may be significant. However, a relatively high dermal LD_{50} value does not necessarily exclude that the compound under some circumstances could be a skin absorption hazard.

The Commission of the European Communities (CEC) has used an LD_{50} of $2 g kg^{-1}$ as a criterion for labelling of chemicals as a skin hazard; to a certain extent, the American Conference of Governmental Industrial Hygienists (ACGIH) has adopted a similar practice for adding a skin denotation to the lists of threshold limit values (TLVs). However, many exceptions to this criterion occur in the TLV list (Scansetti *et al.*, 1988).

Human data

The skin varies in thickness and structure between species. Much information is still lacking with regard to the extent of interspecies differences. Some comparative studies seem to indicate that skin absorption in rats and rabbits is rather high, while that in pigs and monkeys usually approaches skin absorption in humans (Wester and Maibach, 1985a; Wester and Noonan, 1980). However, some of the apparent interspecies differences could be partially due to differences in the site of application, concentration, effects of shaving of the skin, use of occlusion and other experimental conditions. In addition, the interspecies differences have not been uniform with the few substances studied (Wester and Maibach, 1985a; Wester and Noonan, 1980). Thus, the detailed interpretation of experimental animal data must await further validation of the models used.

Most relevant to occupational health are studies of volunteers. Usually, the method of differences is applied. According to the most frequently used method, a hand is placed in the fluid or solution under study or a defined skin area is otherwise exposed for a specified length of time, after which the remaining fluid on the skin is gently wiped off (Wester and Maibach, 1983). The absorption is then measured by assessing the changes in blood concentrations over time (with blood obtained from the other arm), the excretion pattern in urine, or by other relevant methods. The results are then compared with the concentrations resulting from pulmonary absorption of the compound or from an injected dose. Studies of this kind have documented that realistic skin exposures for toluene or xylenes may result in blood concentrations similar to the levels caused by respiratory exposures to the highest permissible air concentration during a working day (Aitio et al., 1984; Engström et al., 1977). However, metabolism within the skin may influence the results. Also, if the compound or its metabolite also occurs endogenously in the body, data obtained from radioactive tracers may be erroneous.

Other researchers have used the direct method for assessing skin penetration (Piotrowski, 1957). In brief, the dissolved compound is applied to the skin area; after the exposure period, the amount of the compound remaining in the solution is determined as an indication of the cutaneous uptake during the exposure period. However, some of the absorbed amount of the compound may be retained in the skin from where it may subsequently evaporate, it may stay bound within the skin or perhaps diffuse through the stratum corneum at a much later time. Thus, this method will invariably result in an overestimation of the penetration rates.

Industrial experience is sometimes useful. For example, biological monitoring data may show whether or not lower absorption is achieved when skin protection is encouraged. More frequently, case reports relate to individual over-exposures where the available information does not allow a quantitative assessment of the significance of percutaneous absorption.

Potentials for classification

Chemical properties

Many factors play a role with regard to skin absorption. Thus, data generated in different laboratories may not necessarily be comparable. Also, one particular test situation will not reflect all cutaneous exposure situations that may occur at the workplace. Although the data base is still rather meagre, it is sufficiently large to allow some limited generalizations concerning the penetration properties of groups of industrial chemicals (Scheuplein and Bronaugh, 1983; Wester and Maibach, 1985). The chemical structure of a compound is important for the potential for skin penetration.

Some small, hydrophilic compounds may follow water through the stratum corneum. However, the diffusivity decreases if the compound possesses one or more polar groups that may interact with polar groups in the stratum corneum. In addition, large molecular size and degree of hydration will also decrease the penetration. Thus, electrolytes applied in aqueous solution do not readily penetrate the skin. Most likely, the ions create a sphere of stable hydration that increases the size of the diffusing unit. Also, the electrical charge of an ion may render it capable of interacting with components of the stratum corneum, thus slowing down the diffusion. For practical purposes, compounds with a molecular weight above 500 will only pass through the skin very slowly, if at all (Schaefer et al., 1982).

Among the polar non-electrolytes, the water-miscible alcohols have been studied as model compounds (figure 1.6) (Scheuplein and Bronaugh, 1983). The penetration of these compounds increases as the lipid solubility increases (methanol constitutes an exception because it is highly caustic). In contrast, addition of another polar group reduces the penetration rate by about 50-fold. Also, the rate decreases if a functional group is substituted by a more polar one. When increasing the length of an aliphatic alcohol, the membrane solubility increases, and thus also the permeability. More than 10 methylene groups in the carbon chain of the alcohol adds little more to the solubility and ultimately reduces the permeability due to the lower mobility of larger molecules. When pure alcohols are applied on the skin, the penetration rates are much lower than the corresponding rates for aqueous solutions. Esters penetrate faster than do the corresponding alcohols, and the unsubstituted hydrocarbons even faster.

With a series of phenol compounds, the highest permeability coefficient was found for 2,4-dichlorophenol and 2,4,6-trichlorophenol, i.e. about 10-fold higher than that of *p*-nitrophenol, while intermediate values were seen with

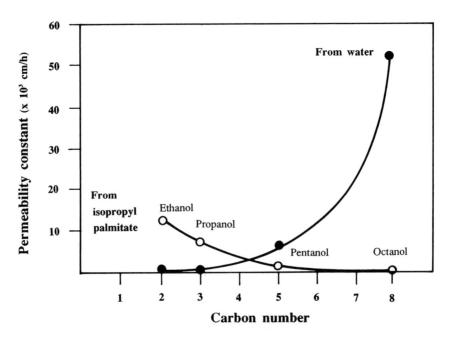

Fig. 1.6. Permeability of alcohols through human skin in vitro. Data from Scheuplein and Bronaugh, 1983.

phenol and cresols (Roberts et al., 1977). Within this series of compounds, a close correlation was obtained between the permeability coefficient and $\log P_{ow}$. A similar correlation was seen with the partition coefficient for water/stratum corneum.

Partition coefficients

In the past, it was often said that lipophilic compounds would more easily pass the skin barrier than would hydrophilic ones (Scheuplein and Bronaugh, 1983). Although the skin has an appreciable lipid content, the statement is true only when the compounds are applied on the skin in a hydrophilic vehicle. The partition coefficient skin-vehicle (K_{MV}) would then result in a higher penetration of the lipophilic compounds due to the lipid solubility of the stratum corneum. However, the reverse may be the case if a lipophilic vehicle is used. For example, N-nitrosodiethanolamine, a water-soluble substance, penetrates the skin 200 times faster from olive oil than from water (Scheuplein and Bronaugh, 1983).

Because the horny layer has both hydrophilic and lipophilic regions, substances which, to some extent, are both water and lipid soluble may more easily penetrate the skin. In fact, a water–ether partition coefficient close to unity seems to promote a better penetration. Also, butanol or a 1:1 mixture of ethanol and water may approximate the solvent character of the stratum corneum. Olive oil, isopropyl palmitate and amyl caproate have been used as analogues for epidermal lipids (Scheuplein and Bronaugh, 1983). These findings are in agreement with Fick's law that emphasizes the importance of the solubility in stratum corneum (K_{MV}). Compounds with both hydrophilic and lipophilic character and of small molecular size tend to penetrate the skin rapidly, occasionally faster than water.

Alternative pathways

For very lipophilic and large molecules, and for some electrolytes, significant skin penetration may occur through alternative pathways. Hair follicles and sweat glands may provide an important shunt for such compounds, especially when diffusion through the horny layer is slow. For example, percutaneous absorption of benzo[a]pyrene, a lipophilic compound that penetrates the skin very slowly, correlated with the hair density in mouse strains; histological evidence suggested that transfollicular diffusion was the main route of penetration (Kao et al., 1988). Additionally, very lipophilic chemicals may conceivably pass slowly through the skin by way of the lipid network in the stratum corneum. The penetration rates are generally very low, but a prolonged absorption may be of significance with very toxic compounds.

Vapours and gases may penetrate the skin if they are sufficiently soluble in the horny layer. For toxic chemicals in the occupational environment, such absorption seems of little significance in comparison to pulmonary absorption of the same airborne concentration. For volatile chemicals, a spill on the skin may cause initial absorption into the outer skin layers, but may later be followed by an almost complete removal by evaporation, unless the skin is occluded.

Structure–activity relationships

Because of the complexities of percutaneous absorption, quantitative structure–activity relationships will not be available for many years. Only gross, qualitative information is available at the present time. Hopefully, as more data are being collected in a structured fashion, better insight will be gained into the factors that govern the rate of percutaneous absorption. However, even when this information becomes available, the specific circumstances will determine the degree of absorption in each skin exposure situation.

The information reviewed in this book is only the tip of the iceberg. Compounds that have shown severe toxicity fortunately become candidates for a ban or restrictions of their use. New compounds are then used instead, but the only information available on percutaneous absorption is often a dermal LD_{50} value. Thus, much of the information on hazards of skin contact is related to compounds of decreasing commercial importance. Therefore, major emphasis must be placed on the development of better guidelines for the prediction of skin absorption hazards.

Sources of variation

The vehicle

Several factors play an important role in determining the rate of skin absorption. Obvious parameters of importance include the size of the skin area involved, the duration of skin contact and the concentration of the compound in the solution. More complex, but often more important factors are the type of vehicle and the integrity of the skin barrier.

Chemicals usually occur as mixtures, and the majority of chemicals encountered in industry are rarely used in pure form. As indicated above, the same compound may have very different absorption rates for the same concentration in, e.g. water, ethanol and oil (Scheuplein and Bronaugh, 1983). An example is shown in figure 1.6. Unfortunately, the significance of various vehicles and other solutes for the penetration rate of a particular chemical has only been explored in a small proportion of the studies published.

A further complicating factor is that prior exposure of the skin to a solvent may increase the permeability of a subsequently applied compound. Thus, the penetrating chemical may not necessarily occur in solution with the enhancing vehicle, but the latter may have previously entered the stratum corneum. Some of these vehicles may even damage the skin, as is discussed below.

Hydration of the skin

The barrier function of human skin is not permanent nor constant. The properties vary with the hydration of the skin. A certain, minimal water content is necessary for a proper barrier function of the horny layer. However, if the hydration of the skin is further increased, the permeability may be augmented by more than 10-fold (Scheuplein, 1978). Hydration can influence the partition coefficient, thickness and diffusivity, and the effect is therefore difficult to predict (Blank, 1985). Limited evidence has shown that postapplication washing may significantly increase the absorption, perhaps due to the increased hydration (Wester and Maibach, 1983). An increase in hydration can also be caused by occlusion of sweaty skin by clothing, ointments or greasy substances on the skin. This factor is important if a chemical is spilled into a rubber boot, or if a chemical penetrates the glove material and thus reaches an occluded surface of the skin. Studies with pharmaceutical agents suggest that occlusion may be the single factor which augments skin penetration the most (Wester and Maibach, 1983). In addition to the increased hydration, occlusion or washing will also result in an increased skin temperature and increased blood flow; these changes will tend to further augment the rate of percutaneous absorption.

Occlusion is of importance in one additional situation. A so-called 'reservoir effect' has been demonstrated for some slowly penetrating substances, in particular the steroids used as topical treatment for dermatological diseases (Scheuplein and Bronaugh, 1983; Wester and Maibach, 1983). Initial skin exposure mainly results in an accumulation in the horny layer. However, long after application and after absorption into the skin, an increased temperature will result in renewed or additional systemic absorption. This 'reservoir effect' may be relevant for slowly diffusing molecules with a large solubility in stratum corneum, if the skin is subsequently occluded by water-impermeable gloves or protective clothing.

Integrity of the skin

A distinct variation in regional skin penetration rates has been documented. Age may also play a role, although the function of stratum corneum seems to be fully preserved in old age. However, the skin of preterm infants may exhibit a permeability 10^2–10^3 times above that seen in full-term infants (Barker *et al.*, 1987). Individual variation must be significant, but has so far received very limited attention.

The skin barrier obviously depends on its integrity for proper functioning. Although no controlled studies are available, dermatological diseases must be considered likely to enhance percutaneous absorption of chemicals. Anecdotal evidence suggests that this consideration is important. For example, a pesticide formulator had higher concentrations of organochlorine pesticides in the blood than other workers similarly exposed; this difference was attributed to the presence of scleroderma in the worker with the high levels (Starr and Clifford, 1971). Hyperproliferation of the stratum corneum, as in psoriasis, exfoliative dermatitis or ichtyosis, result in increased permeability of the skin (Scheuplein and Bronaugh, 1983). The skin of atopic individuals and eczematous areas may also be more permeable. The type of disease and its duration, severity and extent are obviously important factors.

Damaging effects of chemicals

Skin absorption for some compounds may be much more rapid through damaged stratum corneum, i.e. burns, excoriations, dried-out or defatted skin (Scheuplein, 1978; Wester and Maibach, 1983). Delipidization of the skin may be caused by various solvents; some solvents with both polar and non-polar characteristics cause considerable damage to the stratum corneum, perhaps by efficiently extracting both lipids and proteolipids. In particular, a chloroform–methanol mixture (2:1) or ether–ethanol (10:1) have both polar and non-polar characteristics which severely damage the horny layer (Scheuplein, 1978). Subsequent chemical exposure of the skin will most likely result in augmented absorption rates. Skin burns due to solvents can be life-threatening, and removal of damaged skin may be needed to limit systemic uptake of the solvent (Hansbrough et al., 1985).

Many soaps and detergents can damage the skin, and several solvents, such as dimethylsulfoxide (DMSO), cause corrosion, in particular when applied in high concentrations (Scheuplein, 1978). As an indication of the adverse effect on the skin barrier, the penetration rate for N,N-dimethylformamide increases with time (Bortsevich, 1984). The effect on the skin barrier can also be assessed by determining the loss of water vapour induced (van der Valk et al., 1985). To some degree, this damage is reversible, unlike the longer-lasting damage caused by organic solvents. DMSO, N,N-dimethylformamide and N,N-dimethylacetamide are aprotic solvents which differ from water and alcohol in their tendency to accept rather than donate protons; they are useful as solvents for various organic and inorganic substances. Cutaneous permeability is augmented by the resulting displacement of water and denaturation of the horny layer, when rather concentrated amounts of the aprotic solvent are applied to the skin. A derivative of DMSO, $C_{10}MSO$ (C_{10} Methylsulfoxide), is as effective as DMSO at much lower concentrations.

Some anionic and cationic surfactants have the same effect in dilute solution, while non-ionic surfactants appear much less damaging, although they result in a partly reversible decrease of the tissue capacity to bind water. Their effectiveness is due to their unique lipid–polar nature.

Additional factors of possible importance may include the time sequence of repeated skin exposures and the effects of spreading and rubbing of the clothing. A few studies on therapeutic preparations for topical use have indicated the possible significance of such factors (Wester and Maibach, 1983). Again, the occupational health literature is almost devoid of information on these aspects.

Skin denotation

Reasons for a 'skin' denotation

In many countries, compounds considered a skin hazard are identified by a 'skin' denotation on the official list of occupational exposure limits. In general, this practice has the purpose of alerting attention to the fact that cutaneous exposure to these compounds can significantly add to the total systemic exposure when airborne concentrations are kept within the exposure limit.

Only a limited number of compounds with an exposure limit are also regarded a skin hazard. For example, a review of the American TLV list shows that only six inorganic compounds have a 'skin' denotation; aliphatic hydrocarbon derivatives in a liquid state constitute the largest group, and most aromatic compounds are solids; organophosphorus esters and aromatic amines are the largest single groups of chemicals, followed by halogenated hydrocarbons, aliphatic amines and imines, alcohols, nitriles and isocyanates (Scansetti et al., 1988).

The original German approach was to include only chemicals that may contribute significantly to systemic toxicity. Due to the limitations of relevant data and the uncertainties involved, the decision whether or not to consider a compound a skin hazard is difficult. Although the criteria have generally not been disclosed in detail, the practice of 'skin' denotation has been adopted in many countries.

In many countries, special regulations apply to carcinogenic substances. Thus, skin contact with such compounds will be limited, and in that case, no 'skin' denotation may be needed. However, some countries still use 'skin' denotation for carcinogens if they constitute a specific skin absorption hazard. Also, risk of caustic effects on the skin, irritation or allergy may be indicated on product labels, but is apparently also used as a criterion for 'skin' denotation in some countries.

Current practice

Despite the original intentions, many countries have used a 'skin' denotation for chemicals that would be unlikely to cause systemic effects as a result of limited dermal exposures. Unfortunately, the denotation would seem to distinguish good from evil with regard to dermal exposure, while disregarding the numerous shades of grey between the extremes. The discussion above suggests that the majority of chemicals may actually belong within this grey area.

This impression is supported by recent critical reviews of the 'skin' denotation used by ACGIH in the TLV lists (Christensen, 1984; Scansetti et al., 1988). The dermal LD_{50} (below $2\,g\,kg^{-1}$) is the most frequently used criterion, although not uniformly used. For 15% of the compounds, data from humans are given. However, for 20% of the compounds, no justification is given in the documentation (ACGIH, 1986), and for a similar number of compounds, statements are made without proper detail or reference to published data (Scansetti et al., 1988). Other inconsistencies appear from the ACGIH documentation that indicated a skin absorption hazard in 21 cases where the compound had not received a 'skin' denotation on the ACGIH list of exposure limits; the absence of 'skin' denotation was not explained (Christensen, 1984). On the other hand, xylenes were not considered a significant skin hazard in the documentation, while the lists from 1980–1982 did show a 'skin' denotation. However, this denotation has subsequently been removed without explanation, although the wisdom of this decision could be challenged.

International variations

Acording to the legal file at the International Register for Potentially Toxic Chemicals (IRPTC) (1985), the number of chemicals with a 'skin' denotation in the national lists of occupational exposure varies considerably. The list provided by ACGIH includes 179 compounds regarded as skin hazards (and an additional 68 are regulated in this respect in the USA as part of occupational safety and health legislation). However, Switzerland lists as many as 218 compounds (and an additional 57 otherwise), while only 24 chemicals are indicated as a skin hazard on the Hungarian list of exposure limits (and 72 chemicals are otherwise regulated as skin hazards). Within the European Community, the Netherlands list 155 compounds with a 'skin' denotation, while only 74 compounds have been considered a skin hazard in Italy.

Comparison between lists from different countries is difficult because of non-uniform chemical nomenclature. In many cases, more than one chemical name or common name are in use. Further, a compound such as butanol or nitrotoluene may be listed as such or as the individual isomers. Even more difficult are the groups of PCBs and chlorinated naphthalenes where individual congeners and trade names have been used in some lists. To facilitate comparison with exposure lists, common names have been used in this book to identify individual chemicals.

If taken directly from the lists, almost 400 different compounds have been considered a skin hazard in one or more countries. However, the number decreases considerably, when synonyms and isomers are merged, and if all chlorinated biphenyls and all chlorinated naphthalenes are combined. The total number is then reduced to approximately 275 (Grandjean *et al.*, 1988).

Following this revision of the list, figure 1.7 shows the number of chemicals with a 'skin' denotation in official lists of occupational exposure limits in 17 countries. Between 24 and 180 chemicals or groups of chemicals are considered a skin exposure hazard in each country, thus indicating considerable differences.

Partial agreement is apparent from figure 1.7 between the American ACGIH list (143 chemicals) and those of other countries, despite the fact that little more than half of the total number of chemicals considered a skin hazard in any one country is listed as such by ACGIH. In countries with an independent evaluation, such as the Federal Republic of Germany, Sweden and Roumania, between one-fourth and one-third of the chemicals considered a skin hazard are not indicated as such by ACGIH.

As a separate source of information, legislation within the European Community requires that cutaneous hazards be indicated on the label of the compound or product. The decision is based on the dermal LD_{50} value. Several warning phrases may be used. Again, the compounds that must be labelled as a cutaneous hazard differ from the lists used by occupational health authorities. Skin irritation is tested separately and is also included in the labelling (Jacobs *et al.*, 1987).

Whether or not to list a compound as a skin hazard is likely to be influenced by a number of factors, such as existence of skin exposure potentials in each country, significance of other regulations regarding protection of the skin, etc.

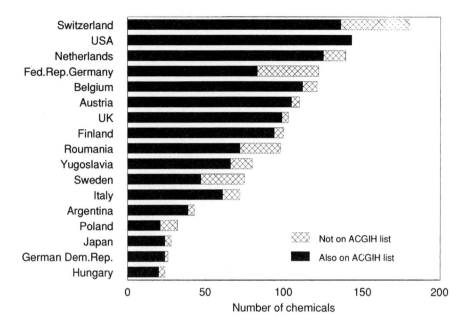

Fig. 1.7. Number of chemicals with a 'skin' denotation on national lists of occupational exposure limits, as compared to the denotations on the ACGIH list. The numbers of chemicals have been adjusted as described in the text.
Data from IRPTC.

Thus, though the ACGIH list and the German list seem to have constituted a basic starting point for many national lists, the differences apparent from figure 1.7 may not be surprising at all.

The disadvantage to the 'skin' denotation is that it represents a gross oversimplification; official lists may tend to reduce the evaluation of skin hazards at the workplace to a mere checking of whether or not a compound has a 'skin' denotation.

Recent modifications

As a uniform method of assigning 'skin' denotations to chemicals, a toxicokinetic approach has been suggested by Fiserova-Bergerova and Pierce (1989). They proposed to calculate percutaneous absorption from an exposure of 2% of the body surface (360 cm^2) to the liquid substance and compare this amount to the absorption from inhalation of the compound at the TLV level. A dermal absorption that corresponded to more than 30% of the pulmonary absorption was considered a criterion for 'skin' denotation. If the TLV is expressed in mg l^{-1}, this criterion will translate to a percutaneous flux (in mg cm^{-2} h^{-1}) that corresponds to the value of $0.75 \times \text{TLV}$.

This standardized approach may be useful in many situations. However, if the TLV is based on irritation rather than systemic toxicity, the calculation will be misleading. Also, for some substances, a contact of a skin area of 360 cm² (corresponding to both hands) with the liquid substance may not be expected to last for a whole working day. However, the calculations allow a useful approximation. The difficulty arises if the criterion of $0.75 \times TLV$ is interpreted as a strict guideline that separates cutaneous hazards from compounds that pose no risk.

Selection of chemicals for review

Validity of 'skin' denotation

Table 1.1 shows the 91 compounds listed as a skin hazard in at least 9 of the 17 countries surveyed by IRPTC. All chemicals were also included on the ACGIH list, except for 1,1-dimethylhydrazine (a 'skin' denotation has now been added by ACGIH). Also, the 'skin' denotation for xylene has recently been removed from the TLV list. The apparent agreement does not indicate that the compounds in table 1.1 are necessarily the most important skin hazards. Rather, many countries have simply copied the TLV list (sometimes including the printing errors).

Following a detailed review of lists used in European countries, three compounds were added (shown in parentheses in the table). In addition, DMSO was added for reasons discussed above, although this compound was only considered a skin hazard in a single country. The 91 chemicals listed in this table appear as a practical selection of those which may most readily be suspected of posing a skin hazard and for which detailed evaluation would seem warranted. Some compounds, such as propargyl alcohol or furfural, may cause considerable damage to the skin, while systemic toxicity is unlikely following cutaneous exposure. Other chemicals, including p-phenylenediamine and methylisocyanate, mainly cause allergic reactions. However, the majority of the chemicals can cause systemic toxicity after percutaneous absorption.

Information on systemic toxicity due to industrial chemicals has recently been compiled in a handbook with a software program (Kimbrough *et al.*, 1989). This data base was consulted and compared to the lists of chemicals with 'skin' denotation. Obviously, when adverse effects in humans have occurred, this fact adds further emphasis to the need for prevention of hazardous exposures. However, only the most obvious chemical hazards are identified this way, because published accounts of human intoxications only represent the tip of the iceberg in relation to toxic chemicals. Thus, even though the chemicals given in italics in table 1.1 have led to documented human toxicity, they should not be regarded the sole possible causes of adverse effects following skin contact.

These chemicals were therefore used as a starting point for detailed evaluation. Primary information on percutaneous absorption and systemic toxicity in relation to cutaneous exposure was retrieved through available data bases and handbooks. However, some chemicals are currently being banned or otherwise phased out in Europe, so that the potential for skin exposure is now minimal or decreasing.

Selection of compounds for detailed review was then performed by using the

Table 1.1 Chemicals with a 'skin' denotation on national list of exposure limits in at least 9 of the 17 countries surveyed, or in the majority of EC countries with independent listings. Additional skin penetrants of apparent significance are shown in parentheses. Compounds with documented systemic effects and known skin exposure potential (italicized) were selected for detailed review and are included in the present report.

Acrylamide	*Hydrazine*
Acrylonitrile	*Hydrogen cyanide*
Aldrin[a]	Lindane[a]
Allyl alcohol	*Malathion*
Aniline	Mercury[b]
Benzene	*Methanol*
bis(2-Chloroethyl) ether[b]	Methyl acrylate[b]
N-Butylamine[b]	N-Methylaniline[b]
(*Carbaryl*)	*Methyl bromide*
Carbon disulfide	2-Methylcyclohexanone[b]
Carbon tetrachloride	*Methyl iodide*
Chlordane[a]	Methyl isocyanate[b]
Chlorinated naphthalenes	Methyl parathion[b]
2-Chloroethanol[b]	Methyl isobutyl carbinol[b]
Chloroprene[b]	Mevinphos[b]
Cresols	Morpholine[b]
Demeton[b]	*Nicotine*
Demeton-methyl[b]	*p-Nitroaniline*
Diazinon	Nitrobenzene
Dichlorvos	p-Nitrochlorobenzene[b]
Dieldrin[a]	*Nitroglycerin*
N,N-Diethyl-2-aminoethanol[b]	*Nitroglycol*
N,N-Dimethylacetamide	Nitrotoluene[b]
N,N-Dimethylaniline[b]	*Paraquat*
N,N-Dimethylformamide	*Parathion*
1,1-Dimethylhydrazine	*PCBs*
Dimethyl sulfate (DMSO)	*Pentachlorophenol*
Dinitrobenzene[b]	*Phenol*
DNOC	p-Phenylenediamine[b]
Dinitrotoluene[b]	*Phenylhydrazine*
1,4-Dioxane	*Picric acid*
Endrin[a]	Propargyl alcohol[b]
Epichlorohydrin	Sodium fluoroacetate[b]
EPN	1,1,2,2-Tetrachloroethane[b]
Ethyl acrylate[b]	*Tetraethyllead*
Ethylene dibromide[b]	Tetramethyllead[b]
Ethylene glycol, monobutyl ether[b]	*Tetryl*
Ethylene glycol, monoethyl ether[b]	*Thallium*
Ethylene glycol, monoethyl ether, acetate[b]	(*Toluene*)
Ethylene glycol, monomethyl ether	o-Toluidine[b]
Ethylene glycol, monomethyl ether, acetate[b]	*Toxaphene*
Ethylene imine[b]	1,1,2-Trichloroethane[b]
Furfural[b]	*2,4,6-Trinitrotoluene*
Heptachlor[a]	*Xylenes*
Hexachloroethane[b]	Xylidines[b]

[a]Skin exposure potential minimal
[b]No information available on systemic toxicity due to skin absorption

following criteria: (i) listed as a skin hazard in at least 9 of the 17 countries surveyed or in the majority of European Community (EC) countries with an independent listing; (ii) shown to cause systemic toxicity in relation to skin absorption; and (iii) known potential for cutaneous exposure at work.

A total of 46 compounds satisfied these criteria, i.e. half of the compounds in the table. Most of the compounds are liquid at room temperature, and many are solvents or pesticides. Three compounds (DMSO, N,N-dimethylacetamide and N,N-dimethylformamide) primarily cause systemic toxicity by making the skin more permeable, thus making penetration easier also for other compounds. Five compounds are known skin hazards, but were left out of the review because they are not in common use any more (e.g. aldrin and chlordane). Several other compounds were reviewed but not included in the report, because the evidence for percutaneous absorption is too limited to allow an evaluation (e.g. chloroprene and demeton). Some were not considered relevant, because they exert their effects on the skin as such and do not cause systemic toxicity when applied on the skin (e.g. furfural and mercury). Finally, some compounds were left out because their main effect is carcinogenic, and all exposures should therefore be minimized (e.g. ethylene dibromide (1,2-dibromoethane) and *o*-toluidine).

This list compares well with the conclusions by Fiserova–Bergerova and Pierce (1989). These authors proposed to include also ethylbenzene, furfural, n-hexane, methyl ethyl ketone (2-butanone), styrene, and trichloroethylene. Of these compounds, only n-hexane and trichloroethylene have caused well documented systemic toxicity in humans (Kimbrough *et al.*, 1989), but the two compounds have received 'skin' denotation only in single countries. The evidence available on these chemicals will be briefly discussed in the appropriate sections of the book. In addition, information has been sought for all 275 compounds with a 'skin' denotation in the countries surveyed (Grandjean *et al.*, 1988); the evidence retrieved is briefly reviewed.

Structure for evaluation

From the above discussion of the scientific evidence on percutaneous absorption, evaluation of skin penetration potentials would appear to be complex and the process is hampered by the lack or incompleteness of available information reported in the scientific literature. With these limitations in mind, the evaluation of possible skin penetrants should consider the following criteria:

(1) Production and use may result in skin exposure (skin contact) with possible percutaneous absorption in workers.
(2) Physicochemical properties of the chemical, as discussed above, suggest that skin penetration is likely to occur to a considerable extent.
(3) Experimental data indicate that skin penetration takes place and may contribute to systemic toxicity.
(4) Case reports on human intoxications and other human observations indicate that skin exposure may be hazardous.

Published information on these topics was identified from personal files, handbooks and data bases, including IRPTC, ECDIN, TOXLINE, RTECS and NIOSHTIC.

Preventive approaches

Planning for prevention

The prevention of skin exposure and percutaneous absorption can be accomplished by several means that are routinely used in occupational health. Important factors are the design of the work process, normal work practices, and general hygiene and safety. Optimal conditions can be obtained by a prudent application of engineering methods, education, labelling and surveillance.

With regard to chemicals that constitute a skin hazard, more specific preventive methods are needed. First of all, such hazards must be identified and evaluated. In the past, safety personnel have usually relied on official lists with 'skin' denotation of hazardous chemicals or warning labels on chemical products. However, as indicated above, evaluation of skin absorption risks are more complicated than that; hopefully, this book will assist in this effort. When such chemicals and products have been identified, any contact with them should preferably be avoided, e.g. by substitution or by encapsulation. If efficient avoidance of contact cannot be achieved, protective gloves or barrier creams can be considered. In deciding on the best preventive approaches, the extent and frequency of skin contact must be taken into account as well as the seriousness of potential adverse health effects.

Treatment

If skin contamination has occurred, the chemical should be removed as quickly as possible. The same applies to drenched clothing. However, removal of the contamination must not result in smearing of the substance on larger areas of the skin. Also, excessive cleansing efforts may be counterproductive, if they result in damage to the skin barrier. Thus, due to the increased hydration, washing with water and soap can significantly increase the absorption of a chemical that has already entered the stratum corneum (Wester and Maibach, 1983, 1985b). For example, washing to remove lindane from the skin immediately after application resulted in an increased absorption (Lange et al., 1981). Local decontamination appears superior to showering (Wester and Maibach, 1985b).

On the basis of such information, Schaefer et al. (1982) recommended that toxic chemicals on the skin should be immediately removed without the use of water, tensides or solvents, and that rubbing should be avoided. Chemical residue on the skin should be removed with a blunt knife, spatula or the like. For lipophilic substances, fat (e.g. butter, margarine or vaseline) should be spread on the contaminated area, left for a brief period and then removed; this treatment should be repeated three times. For hydrophilic substances, the skin area should instead be covered with cellulose powder or a thick paste made with water and silica gel powder, magnesium oxide, charcoal or flour. Stripping of the stratum corneum with adhesive tape should be carried out up to 10 times for very potent substances (the stratum corneum will soon regenerate). The contents of the chemical in the materials used for treatment can subsequently be assessed to determine the effects of the efforts.

Protective clothing

Protective clothing, including gloves, may not necessarily afford full, or even partial protection. As the glove causes occlusion of the skin and thereby increased penetration rates, the safety caused by inappropriate glove or clothing materials may be limited or non-existent. Lack of protective effect has been reported with substances that cause allergic contact dermatitis, e.g. methyl methacrylate (Pegum and Medhurst, 1971) and epoxy resins (Pegum, 1979). Use of protective gloves resulted in a significantly reduced percutaneous uptake of ethylene glycol dinitrate and nitroglycerin (Hogstedt and Ståhl, 1980) and of N,N-dimethylformamide (Lauwerys et al., 1980). On the other hand, methyl parathion may well penetrate through PVC or chloroprene in considerable quantities during a working day (Dedek, 1980).

Testing of glove materials is usually carried out in specially designed test cells. For example, the American Society of Testing and Materials has developed a standard Guide for Test Chemicals to Evaluate Protective Clothing Materials (F-1001-86). Also, a simple instrument has been designed to measure breakthrough times on the shop floor. It consists of a permeation cell and vapour detector that is sensitive to the compound studied. A NIOSH (National Institute for Occupational Safety and Health) Chemical Protective Clothing Portable Test Method is available, as well as some modifications (Berardinelli et al., 1987). The assessment of the breakthrough time will depend on the detection limit of the analytical method. The term penetration in these studies refers to the passage of chemicals through holes and imperfections in the material. In addition to the thickness and intactness of the material, temperature and humidity have important effects on the performance of the material (Bentz, 1987). The breakthrough time has in some instances been less than 10 min (Mellström, 1983), thus indicating very poor performance of the glove material. With most materials, the breakthrough time for most solvents and resins is up to a few hours.

In choosing protective clothing, several factors must be considered (Mellström, 1983). First of all, the degree of protection needed must be compared to the one offered by the materials available. The following information should be sought: breakthrough time, permeation rate and the possible effect of the chemicals on the physical properties of the material. If permeability data cannot be obtained, solubility parameters may be explored; increased permeability occurs when the three-dimensional solubility parameters for the polymer and the penetrant are similar (Hansen and Hansen, 1988). General durability and other mechanical characteristics of the material are also relevant. Components in some glove materials have caused cases of allergic contact dermatitis in the past, and the chemical constituents should therefore be taken into account. Finally, gloves and other protective clothing should fit tightly, interfere as little as possible with normal work and be sufficiently comfortable to be worn continuously. Often, a cotton glove is worn under the protective glove to absorb sweat and to give added protection. At least a dozen types of materials are available, and with new copolymers appearing on the market, the number is increasing. Unfortunately, the documentation for protective effects is unsatisfactory. One of the possible sources of information is DAISY, a data base with the National Institute of Occupational Health in Stockholm, Sweden.

If reliable information on acceptable breakthrough times cannot be obtained, gloves should only be used until they become contaminated. Then they should be replaced by a new pair of gloves.

Barrier creams

Barrier creams would appear as an attractive solution to protection against cutaneous hazards. The ideal barrier cream would fully protect the skin against penetration of a particular chemical for a certain time period, and the cream would itself not cause any adverse effects on the skin.

The effect of barrier creams can be tested by various methods, and animal models appear advantageous (Boman *et al.*, 1982). However, experimental support for the protective properties is still very limited, and creams have in some cases promoted the development of contact dermatitis (Mellström, 1983). Some protective effects have been documented with regard to percutaneous absorption of N,N-dimethylformamide (Lauwerys *et al.*, 1980) and toluene (Guillemin *et al.*, 1974), but not with m-xylene (Lauwerys *et al.*, 1978). However, a normal hand cream offered a slightly larger reduction in skin absorption of 1,1,1-trichloroethane than two barrier creams, while treatment of damaged skin with a barrier cream even increased the absorption of 1-butanol (Boman, 1989).

When indicated by valid documentation, barrier creams may be used when only a minimal protection is needed and protective gloves would hinder normal work. Such creams may also help focus attention on the importance of dermal hygiene, and they may make it easier to remove contamination from the skin.

Both protective clothing and barrier creams offer only temporary protection.

References

AITO, A., PEKARI, K. and JÄRVISALO, J. (1984), Skin absorption as a source of error in biological monitoring, *Scandinavian Journal of Work and Environmental Health*, **10**, 317-20.

ACGIH (1986), *Documentation of the threshold limit values and biological exposure indices*, 5th edn (Cincinnati: American Conference of Governmental Industrial Hygienists).

ANDERSON, J. E., HOFFMAN, S. J. and PETERS, C. R. (1972), Factors influencing reverse osmosis rejection of organic solutes from aqueous solution, *Journal of Physical Chemistry*, **76**, 4006-11.

BARKER, N., HADGRAFT, J. and RUTTER, N. (1987), Skin permeability in the newborn, *Journal of Investigative Dermatology*, **88**, 409-11.

BENTZ, A. P. (1987), U.S. Coast Guard's protective clothing material test program. In *Second Scandinavian Symposium on Protective Clothing against Chemicals and other Health Risks*, eds G. Mellström and B. Carlsson (Arbete och Hälsa 1987, **12**, pp. 51-60) (Stockholm: National Board of Occupational Safety and Health).

BERARDINELLI, S. P., RUSCZEK, R. A. and MICKELSEN, R. L. (1987), A portable chemical protective clothing test method: Application at a chemical plant, *American Industrial Hygiene Association Journal*, **48**, 804-8.

BLANK, I. H. (1985), The effect of hydration on the permeability of the skin. In *Percutaneous absorption*, eds R. L. Bronaugh and H. I. Maibach, pp. 97-105 (New York: Marcel Dekker).

BOMAN, A. (1989), *Factors influencing the percutaneous absorption of organic solvents*, (Arbete och Hälsa Vol. 1989: 11) (Stockholm: National Institute of Occupational Health).

BOMAN, A., WAHLBERG, J. E. and JOHANSSON, G. (1982), A method for the study of the effect of barrier creams and protective gloves on the percutaneous absorption of solvents, *Dermatology*, **164**, 157-60.

BORTSEVICH, S. V. (1984), The problem of the hygienic importance of dimethylformamide absorption through the skin (in Russian), *Gigiena Truda I Professionalnyn Zabolevaniia*, **11**, 55-7.

BRONAUGH, R. L. and MAIBACH, H. I. (1985), In vitro models for human percutaneous absorption. In *Models in Dermatology*, eds H. I. Maibach and N. J. Lowe, Vol. 2, pp. 178-88 (Basel: Karger).

CARVER, M. P., WILLIAMS, P. L. and RIVIERE, J. E. (1989), The isolated perfused porcine skin flap III. Percutaneous absorption pharmacokinetics of organophosphates, steroids, benzoic acid, and caffeine, *Toxicology and Applied Pharmacology*, **97**, 324-37.

CHRISTENSEN, U. L. (1984), *Skin permeability*, (in Danish) (Institute for Occupational Environment, Technical University, Lyngby).

DEDEK, W. (1980), Solubility factors affecting pesticide penetration through skin and protective clothing. In *Field Worker exposure during pesticide application*, eds W. F. Tordoir and E. A. H. Heemstrahequin. (*Studies in Environmental Science*, Vol. 7, pp. 47-50.) (Amsterdam: Elsevier).

DUPUIS, D., ROUGIER, A., ROGUET, R., LOTTE, C. and KALOPISSIS, G. (1984), In vivo relationship between horny layer reservoir effect and percutaneous absorption in human and rat, *Journal of Investigative Dermatology*, **82**, 353-6.

DUPUIS, D., ROUGIER, A., ROGUET, R. and LOTTE, C. (1986), The measurement of the stratum corneum reservoir: a simple method to predict the influence of vehicles on in vivo percutaneous absorption, *British Journal of Dermatology*, **115**, 233-8.

ENGSTRÖM, K., HUSMAN, K. and RIIHIMÄKI, V. (1977), Percutaneous absorption of m-xylene in man, *International Archives of Occupational and Environmental Health*, **39**, 181-9.

FENSKE, R. A. (1988), Correlation of fluorescent tracer measurements of dermal exposure and urinary metabolite excretion during occupational exposure to malathion, *American Industrial Hygiene Association Journal*, **49**, 438-44.

FISEROVA-BERGEROVA, V. and PIERCE, J. T. (1989), Biological monitoring V. Dermal absorption, *Applications of Industrial Hygiene*, **4**, F14-F21.

FRANZ, T. J. (1975), Percutaneous absorption. On the relevance of in vitro data, *Journal of Investigative Dermatology*, **64**, 190-5.

FREDERIKSSON, T., FARRIOR, W. L. Jr and WITTER, R. F. (1961), Studies of the percutaneous absorption of parathion and paraoxon-I. Hydrolysis and metabolism within the skin, *Acta Dermato-Venereologica*, **41**, 335-43.

GARLAND, T. O. and PATTERSON, M. W. H. (1967), Six cases of acrylamide poisoning, *British Medical Journal*, **4**, 134-8.

GARY-BOBO, C. M., DIPOLO, R. and SOLOMON, A. K. (1969), Role of hydrogenbonding in nonelectrolyte diffusion through dense artificial membranes, *Journal of General Physiology*, **54**, 369-82.
GRANDJEAN, P., BERLIN, A., GILBERT, M. and PENNING, W. (1988), Preventing percutaneous absorption of industrial chemicals: The 'skin' denotation, *American Journal of Industrial Medicine*, **14**, 97-107.
GRASSO, P. and LANSDOWN, A. B. G. (1972), Methods of measuring, and factors affecting, percutaneous absorption, *Journal of the Society of Cosmetic Chemistry*, **23**, 481-521.
GUILLEMIN, M., MARSET, J. C., LOB, M. and RIGUETZ, J. (1974), Simple method to determine the efficiency of a cream used for skin protection against solvents, *British Journal of Industrial Medicine*, **31**, 310-6.
GUY, R. H. and HADGRAFT, J. (1985), Mathematical models in percutaneous absorption. In *Models in Dermatology*, eds H. I. Maibach and N. J. Lowe, Vol. 2, pp. 170-77 (Basel: Karger).
HANSBROUGH, J. F., ZPATA-SIRVENT, R., DOMINIC, W., SULLIVAN, J., BOSWICK, J. and WANG, X.-W. (1985), Hydrocarbon contact injuries, *Journal of Trauma*, 250-2.
HANSCH, C. and LEO, A. (1979), *Substituent constants for correlation analysis in chemistry and biology* (New York: Wiley).
HANSCH, C., KIM, D., LEO, A. J., NOVELLINO, E., SILIPO, C. and VITTORIA, A. (1989), Toward a quantitative comparative toxicology of organic compounds, *CRC Critical Review of Toxicology*, **19**, 185-226.
HANSEN, C. M. and ANDERSEN, B. H. (1988), The affinities of organic solvents in biological systems, *American Industrial Hygiene Association Journal*, **49**, 301-8.
HANSEN, C. M. and HANSEN, K. M. (1988), Solubility parameter prediction of the barrier properties of chemical protective clothing. In *Performance of Protective Clothing*, Second Symposium, eds S. Z. Mansdorf, R. Sager and A. P. Nielsen, Special Technical Publ. 989, pp. 197-208 (Philadelphia, PA: American Society for Testing and Materials).
HOGSTEDT, C. and STÅHL, R. (1980), Skin absorption and protective gloves in dynamite work, *American Industrial Hygiene Association Journal*, **41**, 367-72.
IRPTC (1985), *International Register of Potentially Toxic Chemicals*, Part A. (Geneva: United Nations Environment Programme).
JACOBS, G., MARTENS, M. and MOSSELMANS, G. (1987), Proposal of limit concentrations for skin irritation within the context of a new EEC directive on the classification and labelling of preparations, *Regulatory Pharmacology and Toxicology*, **7**, 379-80.
JEGIER, Z. (1974), Health hazards in insecticide spraying of crops, *Archives of Environmental Health*, **8**, 670-4.
KAO, J., HALL, J. and HELMAN, G. (1988), In vitro percutaneous absorption in mouse skin: influence of skin appendages, *Toxicology and Applied Pharmacology*, **94**, 93-103.
KAZEN, C., BLOOMER, A., WELCH, R., OUDBIER, A. and PRICE, H. (1974), Persistence of pesticides on the hands of some occupationally exposed people, *Archives of Environmental Health*, **29**, 315-8.
KIMBROUGH, R. D., MAHAFFEY, K. R., GRANDJEAN, P., SANDOE, S.-H. and RUTSTEIN, D. D. (1989), *Clinical effects of environmental chemicals* (New York: Hemisphere).
LANGE, M., NITZSCHE, K. and ZESCH, A. (1981), Percutaneous absorption of lindane in healthy volunteers and scabies patients, *Archives of Dermatological Research*, **271**, 387-99.
LAUWERYS, R. R., DATH, T., LACHAPELLE, J.-M., BUCHET, J.-P. and ROELS, H. (1978), The influence of two barrier creams on the percutaneous absorption of *m*-xylene in man, *Journal of Occupational Medicine*, **20**, 17-20.

LAUWERYS, R. R., KIVITS, A., LHOIR, M., RIGOLET, P., HOUBEAU, D., BUCHET, J. P. and ROELS, H. A. (1980), Biological surveillance of workers exposed to dimethylformamide and the influence of skin protection on its percutaneous absorption, *International Archives of Occupational and Environmental Health*, **45**, 189-203.

LEES, P. S. J., CORN, M. and BREYSSE, P. N. (1987), Evidence for dermal absorption as the major route of body entry during exposure of transformer maintenance and repairmen to PCBs, *American Industrial Hygiene Association Journal*, **48**, 257-64.

LEO, A., HANSCH, C. and ELKINS, D. (1971), Partition coefficients and their uses, *Chemical Reviews*, **71**, 525-616.

MAIBACH, H. I., FELDMANN, R. J., MILBY, T. H. and SERAT, W. F. (1971), Regional variation in percutaneous penetration in man, pesticides, *Archives of Environmental Health*, **23**, 208-11.

MELLSTRÖM, G. (1983), Protective gloves and barrier creams (in Swedish). (Undersökningsrapport 1983, pp. 28-30) (Stockholm: National Board of Occupational Safety and Health).

NATER, J. P. and DE GROOT, A. C. (1983), Systemic side effects caused by topically applied drugs. In *Unwanted Effects of Cosmetics and Drugs Used in Dermatology*, Chap. 16, pp. 140-72 (Amsterdam: Excerpta Medica).

NOONAN, P. K. and WESTER, R. C. (1985), Cutaneous metabolism of xenobiotics. In *Percutaneous absorption*, eds R. L. Bronaugh and H. I. Maibach, pp. 65-85 (New York: Marcel Dekker).

PEGUM, J. S. (1979), Penetration of protective gloves by epoxy resin, *Contact Dermatitis*, **5**, 281-3.

PEGUM, J. S. and MEDHURST, F. A. (1971), Contact dermatitis from penetration of rubber gloves by acrylic monomer, *British Medical Journal*, **2**, 141-3.

PIOTROWSKI, J. (1957), Quantitative estimation of aniline absorption through the skin in man, *Journal of Hygiene, Epidemiology, Microbiology and Immunology*, **1**, 23-32.

RIVIERE, J. E., BOWMAN, K. F., MONTEIRO-RIVIERE, N. A., DIX, L. P. and CARVER, M. P. (1986), The isolated perfused porcine skin flap (IPPSF) I. A novel in vitro model for percutaneous absorption and cutaneous toxicology studies, *Fundamental Applications of Toxicology*, **7**, 444-53.

ROBERTS, M. S., ANDERSON, R. A. and SWARBRICK, J. (1977), Permeability of human epidermis to phenolic compounds, *Journal of Pharmacy and Pharmacology*, **29**, 677-83.

ROUGIER, A., DUPUIS, D., LOTTE, C., ROGUET, R. and SCHAEFER, H. (1983), In vivo correlation between stratum corneum reservoir function and percutaneous absorption, *Journal of Investigative Dermatology*, **81**, 275-8.

ROUGIER, A., DUPUIS, D., LOTTE, C. and ROGUET, R. (1985), The measurement of the stratum corneum reservoir, a predictive method for in vivo percutaneous absorption studies: Influence of application time, *Journal of Investigative Dermatology*, **84**, 66-8.

ROUGIER, A., DUPUIS, D., LOTTE, C., ROGUET, R., WESTER, R. C. and MAIBACH, H. I. (1986), Regional variation in percutaneous absorption in man: measurement by the stripping method, *Archives of Dermatological Research*, **278**, 465-9.

SCANSETTI, G., PIOLATTO, G. and RUBINO, G. F. (1988), Skin notation in the context of workplace exposure standards, *American Journal of Industrial Medicine*, **14**, 725-32.

SCHAEFER, H., ZESCH, A. and STÜTTGEN, G. (1982), *Skin permeability* (Berlin: Springer Verlag).

SCHEUPLEIN, R. (1978), Site variations in diffusion and permeability. In *The Physiology and Pathophysiology of the Skin*, ed. A. Jarrett, Vol. 5, pp. 1731-52 (London: Academic Press).

SCHEUPLEIN, R. (1978), Skin permeation. In *The Physiology and Pathophysiology of the Skin*, ed. A. Jarrett, Vol. 5, pp. 1693-730 (London: Academic Press).
SCHEUPLEIN, R. (1978), The skin as a barrier. In *The Physiology and Pathophysiology of the Skin*, ed. A. Jarrett, Vol. 5, pp. 1669-92 (London: Academic Press).
SCHEUPLEIN, R. J. and BLANK, I. H. (1971), Permeability of the skin, *Physiology Review*, 51, 702-47.
SCHEUPLEIN, R. J. and BRONAUGH, R. L. (1983), Percutaneous absorption. In *Biochemistry and Physiology of the Skin*, ed. L. A. Goldsmith, pp. 1255-95 (Oxford: Oxford University Press).
STARR, H. G. and CLIFFORD, N. J. (1971), Absorption of pesticides in a chronic skin disease, *Archives of Environmental Health*, 22, 396-400.
TSURUTA, H. (1975), Percutaneous absorption of organic solvents. I. Comparative study of the in vivo percutaneous absorption of chlorinated solvents in mice, *Industrial Health*, 13, 227-36.
TSURUTA, H. (1982), Percutaneous absorption of organic solvents. III. On the penetration rates of hydrophobic solvents through excised rat skin, *Industrial Health*, 20, 335-45.
VAN HOOIDONK, C., CENLEN, B. I., KIENHUIS, H. and BOCK, J. (1980), Rate of skin penetration of organophosphates measured in diffusion cells. In *Mechanism of Toxicity and Hazard Evaluation*, eds B. Holmstedt, R. Lauwerys, M. Mercier and M. Roberfroid, pp. 643-6 (Amsterdam: Elsevier).
VAN DER VALK, P. G. M., NATER, J. P. and BLEUMINK, E. (1985), The influence of low concentrations of irritants on skin barrier function as determined by water vapour loss, *Dermatosen*, 33, 89-91.
WESTER, R. C. and MAIBACH, H. I. (1983), Cutaneous pharmacokinetics: 10 steps to percutaneous absorption, *Drug Metabolism Review*, 14, 169-205.
WESTER, R. C. and MAIBACH, H. I. (1984), Advances in percutaneous absorption. In *Cutaneous Toxicity*, eds V. A. Drill and P. Lazar, pp. 29-40 (New York: Raven).
WESTER, R. C. and MAIBACH, H. I. (1985a), Animal models for percutaneous absorption. In *Models in Dermatology*, eds H. I. Maibach and N. J. Lowe, Vol. 2, pp. 159-69 (Basel: Karger).
WESTER, R. C. and MAIBACH, H. I. (1985b), Dermal decontamination and percutaneous absorption. In *Percutaneous absorption*, eds R. L. Bronaugh and H. I. Maibach, pp. 327-33 (New York: Marcel Dekker).
WESTER, R. C. and MAIBACH, H. I. (1985c), Structure-activity correlations in percutaneous absorption. In *Percutaneous absorption*, eds R. L. Bronaugh and H. I. Maibach, pp. 107-23 (New York: Marcel Dekker).
WESTER, R. C., MOBAYEN, M. and MAIBACH, H. I. (1987), In vivo and in vitro absorption and binding to powdered stratum corneum as methods to evaluate skin absorption of environmental chemical contaminants from ground and surface water, *Journal of Toxicological and Environmental Health*, 21, 367-74.
WESTER, R. C. and NOONAN, P. K. (1980), Relevance of animal models for percutaneous absorption, *International Journal of Pharmaceutics*, 7, 99-110.
WHO (1982), *Recommended health-based limits in occupational exposure to pesticides*. Report of a WHO Study Group, (Technical Report Series 677) (Geneva: World Health Organization).
WHO (World Health Organization) (1986), Field surveys of exposure to pesticides standard protocol, *Toxicology Letters*, 33, 223-35.
WOJCIECHOWSKI, Z., PERSHING, L. K., HUETHER, S., LEONARD, L., BURTON, S. A., HIGUCHI, W. I. and KRUEGER, G. G. (1987), An experimental skin sandwich flap on an independent vascular supply for the study of percutaneous absorption, *Journal of Investigative Dermatology*, 88, 439-46.
WOLFE, H. R., DURHAM, W. F. and ARMSTRONG, J. F. (1967), Exposure of workers to pesticides, *Archives of Environmental Health*, 14, 622-33.

WOLFF, M. S., HERBERT, R., MARCUS, M., RIVERA, M., LANDRIGAN, P. J. and ANDREWS, L. R. (1989), Polycyclic aromatic hydrocarbon (PAH) residues on skin in relation to air levels among roofers, *Archives of Environmental Health*, **44**, 157-63.

WURSTER, D. E. (1978), Some physical-chemical factors influencing percutaneous absorption from dermatologicals, *Current Problems in Dermatology*, **7**, 156-71.

Section B
Chemicals as percutaneous hazards

# Chapter 2
Inorganic and organometal compounds

Introduction

The inorganic and organometal compounds most frequently considered to be a skin absorption hazard are indicated in table 2.1. The most important of these are considered separately below.

Electrolytes applied in aqueous solution do not readily penetrate the skin. The hydrated ions constitute a diffusing unit of relatively large size that will slow down the penetration rate. Also, the electrical charge of an ion may render it capable of interacting with components of the stratum corneum. Accordingly, percutaneous absorption of soluble metal compounds in guinea pigs was very slow for cobalt chloride, zinc chloride and silver nitrate, while some mercury compounds penetrated somewhat faster (Skog and Wahlberg, 1964). However, copper intoxication has been recorded after therapeutic application of a copper sulfate solution to burns (Holtzman *et al.*, 1966). Also, an explosion of copper azide caused impaction of copper particles in the skin and caused slow, prolonged systemic absorption of the metal (Bentur *et al.*, 1988). Such cases must be considered exceptional, but they serve to illustrate that serious percutaneous absorption can occur under the right conditions, even with compounds that are of low toxicity and limited percutaneous absorption rate. That thallium has received wide 'skin' denotation is due to its persistence in the body and high chronic toxicity. With mercury, opinions appear to vary. In the USA, only elemental

Table 2.1 Inorganic and organometal compounds that are considered a skin hazard in several countries (y = yes, n = no, o = other regulation). Those that are considered in detail are italicized.*

CAS No.	Chemical	Number of countries	FRG	Sweden	USA
17702-41-9	Decaborane	8	y	n	y
74-90-8	*Hydrogen cyanide*	13	y	y	y
	Cyanides	7	y	y	y
78-00-2	*Tetraethyllead*	13	y	y	y
75-74-1	Tetramethyllead	8	y	y	y
62-74-8	Sodium fluoroacetate	9	y	n	y
1189-85-1	*tert*-Butyl chromate	7	n	n	y
	Mercury (all)	11	y	y	y
	Thallium	9	o	n	y

*Hydrazine belongs to this group of compounds but has been considered in a separate chapter with other nitrogen compounds.

mercury has a 'skin' denotation, while only organic mercury compounds are considered a skin hazard in the FRG and Sweden.

Inorganic compounds listed in at least one country include sodium tetraborate, chromic acid, and chromium oxide. These compounds can cause caustic or allergenic effects (Fisher, 1986), and that is probably the basis for their 'skin' denotation.

Arsenic(III) chloride has been listed because of its high toxicity. Surprisingly, other arsenic compounds have not been listed as a skin absorption hazard in any of the countries surveyed.

Other small, hydrophilic molecules may pass through the stratum corneum more easily if they are without polar groups, e.g. metal salts or organic acids that do not dissociate at the pH in the skin. Thus, lead salts of hydrophobic organic acids, such as lead naphthenate, would be expected to penetrate the skin. However, such compounds have not been found to be listed separately. Other metal compounds, such as cadmium diethyldithiocarbamate which is used as a rubber additive, could potentially penetrate the skin in significant amounts.

Organometal compounds have a covalent bond between the metal and carbon; they are usually lipophilic, but organometal salts tend to be amphiphilic, i.e. both hydrophilic and lipophilic. Skin penetration of toxic quantities of tetraethyllead is well established, and structural similarity would suggest that tetramethyllead should be considered a skin hazard as well. However, the evidence available in this area is insufficient, and the latter compound may be considerably less toxic (Grandjean, 1984).

In addition to the tetraalkylleads, recognized skin hazards include organotin compounds; in some countries specific compounds are indicated, i.e. cyhexatin (tricyclohexyltin hydroxide), dibutyltin dichloride, and triphenyltin acetate or hydroxide. Similarly, all organomercury compounds are frequently considered a cutaneous hazard, but ethylmercury chloride or phosphate, methoxyethylmercury acetate, and methylmercury are also listed separately. Two manganese compounds used as petrol additives (manganesecyclopentadienyltricarbonyl and 2-methylcyclopentadienylmanganesetricarbonyl) are included in exposure lists in some countries, including the USA, and have a 'skin' denotation. Nickel carbonyl has been listed, e.g. in the FRG, but because of its high vapour pressure at 20°C this organometal compound can hardly be considered an important skin hazard.

Hydrogen sulfide in aqueous solution may penetrate the skin several-fold faster than water (Schaefer et al., 1982), and it also acts as a skin irritant. Other hydrides with 'skin' denotation are arsine, diborane, and pentaborane. These compounds are primarily inhalation hazards.

Several cyanide-related compounds have occasionally received 'skin' denotation, including calcium cyanamide, cyanogen (ethane dinitrile), and sodium cyanate. They must be considered a much lesser percutaneous absorption hazard than hydrogen cyanide.

Hydrogen cyanide and cyanide salts

Skin exposure potential

Hydrogen cyanide is primarily produced from methane and ammonia, and the major part of the production is used as part of a larger process system for synthetic fibres, plastics etc., but some is used for fumigation, electroplating and other purposes where skin contact may occur (NIOSH, 1976). Both the sodium and calcium cyanides have wide applications in pesticide production, metallurgical processes and laboratory procedures (NIOSH, 1976). In the USA, an estimated 1000 workers are exposed to hydrogen cyanide, 20 000 to sodium cyanide (NIOSH, 1976).

Physicochemical properties

Hydrogen cyanide is a liquid at room temperature, but with a boiling point of 26°C, evaporation may be considerable. It is very soluble in water, ethanol and ether. The reported octanol/water partition coefficient ($\log P_{ow}$) for hydrogen cyanide is given as 0.35 or 1.07. Cyanide salts are solids at room temperature. Sodium and potassium cyanide are easily soluble in water. Calcium cyanide decomposes and forms calcium hydroxide and hydrogen cyanide; other cyanides are generally insoluble and therefore constitute a minimal skin absorption hazard. These compounds are all of small molecular size, and hydrogen cyanide is slightly lipidilic.

Experimental data

Early experiments by Walton and Witherspoon (1926) suggested that hydrogen cyanide vapour may penetrate the skin of dogs and guinea pigs, and concentrations of several thousand p.p.m. proved fatal. The permeability constant k_p for cyanide in aqueous solution applied to human skin *in vitro* was $3.5 \times 10^4 \text{ cm h}^{-1}$, and the lag time was 1.5 h; with hydrogen cyanide in water, k_p was much higher, i.e. $100 \times 10^4 \text{ cm h}^{-1}$, and the lag time only 3 min (Dugard and Scott, 1984). LD_{50} values for rabbits following percutaneous exposure to aqueous solutions of cyanides were 6.8, 7.7 and 8.9 mg kg^{-1} for hydrogen, sodium and potassium cyanide, respectively; these values were about 10-fold above LD_{50} levels after injection of the cyanides (Ballantyne, 1983).

Human data

Several early cases of poisoning resulted from contact with cyanide solutions, although inhalation may in some cases have contributed to the exposure (NIOSH, 1976). Liquid hydrogen cyanide ran over the bare hand of a worker who wore a respirator, and within 5 min he collapsed, thus suggesting rapid percutaneous absorption (Potter, 1950). Tovo (1955) described a fatal case possibly due to percutaneous absorption of potassium cyanide. Unfortunately, the exact penetration rate and relative contribution by skin absorption to total uptake are hard to judge from the evidence available. Cyanide prevents the uptake of oxygen by the body tissues and results in asphyxia. The clinical picture may include metabolic acidosis and pulmonary oedema (Graham et al., 1977). Although cases of possible chronic poisoning have been recorded in industry (NIOSH, 1976), acute toxicity is the primary issue.

Conclusions

Hydrogen cyanide and soluble cyanide salts may penetrate the skin, although no detailed data on the actual rates of penetration are available. Due to the considerable acute toxicity of cyanides, they should be regarded a skin exposure hazard.

Tetraethyllead

Skin exposure potential

The tetraethyllead (tetraethylplumbane) production in 1980 within the EC was 41 000 tonnes and this was mostly consumed in EC countries. More recently, the production and use have declined. The predominant use of this substance is as an octane-booster in petrol, but small quantities may be used for catalytic or pesticidal purposes. Tetraethyllead is usually mixed with tetramethyllead (tetramethylplumbane), the so-called scavengers ethylene dibromide and ethylene dichloride and a dye. Such mixtures are often referred to as ethyl fluid and are marketed under various trade names, according to its manufacturer and composition. The ethyl fluid is transported to the refineries where it is added to petrol to achieve the lead concentration desired. Cutaneous exposures may especially occur at production facilities and at refineries where the most hazardous procedure appears to be the cleaning of storage tanks. Small concentrations of this chemical are present in all grades of leaded petrol.

Physicochemical properties

Tetraethyllead is a liquid at room temperature and has a vapour pressure of 0.3 mm Hg at 20°C. It is very lipid soluble, and 1 l of water will only dissolve 0.25 mg of this compound, while it is easily soluble in ethanol and acetone.

Experimental data

Dermal LD_{50} levels for tetraethyllead are 0.7–1.5 g kg^{-1} in rabbits and about the same for dogs and guinea pigs; when tetraethyllead is given orally or intraperitoneally the LD_{50} values vary from a low 15 mg kg^{-1} (rat) to a high 100 mg kg^{-1} (rabbit) (Seawright, 1984). However, evaporation of tetraethyllead could limit the amount absorbed under some of the experimental conditions. Unfortunately, no recent experimental evidence is available on skin penetration. Early studies using experimental animals showed that the absorption of tetraethyllead was slow, and penetration did not reach a maximum until 4–6 h after application of the compound (Kehoe, 1927), but lethal quantities may be absorbed within an hour or so (Kehoe and Thamann, 1931). When the pure tetraethyllead was applied to the skin without occlusion, only about 5% was absorbed while the remainder evaporated (Laug and Kunze, 1948). When this compound is diluted with petrol, the lipophilicity of tetraethyllead would prevent it from penetrating very rapidly into the horny layer of the skin, in which it is probably less soluble. Thus, one study indicated more than 50 years ago that a 0.1% solution of tetraethyllead in petrol would result in insignificant absorption (Kehoe, 1931). However, if the petrol evaporates and leaves some tetraethyllead on the skin, additional skin absorption could take place.

Human data

About 150 fatal cases of organolead poisoning have been described in the scientific literature, and a total of about 1000 intoxications have been reported (Grandjean, 1984). In the majority, tetraethyllead was the compound responsible for the toxicity. Skin absorption was of importance in several cases where primitive hygiene conditions prevailed and insufficient protective clothing was provided. For example, one male employee at a production plant was splashed with tetraethyllead on head and thighs and died from the intoxication (Corsi and Picotti, 1965). Although some information on urinary lead excretion is available, the exact contribution from skin absorption is difficult to evaluate from the human exposure data. The lack of correlation between air levels of tetraethyllead and urinary lead excretion in exposed workers is believed to be due to cutaneous absorption of tetraethyllead (American Conference of Governmental Industrial Hygienists, ACGIH, 1986). After an acute exposure to a high dose of tetraethyllead, symptoms do not develop for some time, probably because dealkylation to triethyllead, which is responsible for the toxicity, has to take place (Grandjean, 1984). The first non-specific symptoms may occur one or more hours after the exposure and usually include weakness, fatigue, headache and nausea. Then follow insomnia, anxiety, ataxia, tremor and other indications of central nervous system toxicity. In chronic exposure cases, symptoms similar to the non-specific symptoms of the prodromal period have frequently been reported, but too little is known about the effects of long-term exposures. In addition to neurotoxicity, adverse effects on liver enzymes and chromosomes may be of concern. The lethal amount may be of the order of magnitude of 1 g, but the absorption of milligram quantities on a daily basis would invariably lead to accumulation and toxicity.

Conclusion

Tetraethyllead may penetrate the skin in significant quantities, and dermal absorption has added to the dosage encountered in numerous cases of intoxications.

Thallium

Skin exposure potential

Thallium is primarily obtained from lead-chamber sludge by precipitation with zinc. Thallium and thallium salts have a variety of uses, including production of optical-quality glass, rodenticides (mainly in the past), and photographic formulations, but the number of exposed workers is probably small.

Physicochemical properties

Many thallium salts (including the sulfide) are sparingly soluble, but thallic chloride, thallium cyanide, thallium fluoride, thallous hydroxide and thallium nitrite are easily soluble. As with potassium, the thallium(I) ion would tend to accumulate intracellularly. The rate of skin absorption will be limited by binding to sulfhydryl groups and by the size of the hydrated ion.

Experimental data

The LD_{50} for thallous carbonate after cutaneous application in the rat is 117 mg kg^{-1} (RTECS, 1983), i.e. several-fold higher than those obtained by other administration routes. The rate of penetration of a 2% thallous carbonate solution rubbed at the tail skin of mice was 0.015 mg cm^{-2} h^{-1}, half of the animals died within 1 h and all of them within 3 h (International Register of Potentially Toxic Chemicals, IRPTC, 1982). Although confirmatory evidence is lacking, the data seem to document beyond doubt that percutaneous absorption may occur.

Human data

Thallium acetate has been used for depilatory plasters in the treatment of ringworm; following such treatment, urinary excretion is very variable (Carson and Smith, 1977). The symptoms resulting from percutaneous absorption are similar to those seen after ingestion or inhalation: gastroenteritis, muscle pain, kidney damage and skin eruptions (Prick *et al.*, 1955). More serious effects may involve almost all organ systems and include delirium, convulsions and cardiotoxicity; polyneuritis and alopecia occur at least 10 days after an acute exposure, and a protracted course may be seen in chronic exposures (Carson and Smith, 1977). Cases of occupational thallium poisoning have been described where skin absorption was apparently the cause (Glömme and Sjöström, 1955; Richeson, 1958).

Conclusions

Limited evidence supports the notion that significant absorption may occur through the skin. Due to the considerable and cumulative toxicity of thallium, even a slow percutaneous uptake may be of concern. Thallium and its soluble salts should therefore be regarded a skin hazard.

Other inorganic compounds

Mercury

In the past, mercury and mercury chloride were used in ointments, often against freckles. Such treatments were shown to cause percutaneous absorption of mercury in potentially fatal amounts (Schamberg et al., 1918). Several more recent studies confirmed this finding and documented the importance of the vehicle, type of compound and the concentration. In guinea pigs, up to 3.2–4.5% of mercuric chloride, potassium mercuric iodide and methyl mercury dicyandiamide applied to the skin was absorbed within 5 h (Skog and Wahlberg, 1964). Maximal absorption in guinea pigs was seen with a concentration of 16 mg (Hg)/ml, and higher concentrations were thought to cause a denaturation of the skin, thus inhibiting absorption of the mercury (Friberg et al., 1961). Topical application of mercury compounds and excessive skin contamination with mercury at work must be considered a rarity today. However, dermal application of an antilice formula containing metallic mercury and fats recently resulted in the development of erythema, signs of systemic mercury poisoning and excessive urinary mercury excretion; although respiratory uptake of mercury could not be ruled out, percutaneous absorption seemed the most important route of absorption (Bourgeois et al., 1986). Mercury chloride (calomel) was, particularly in the past, a frequent cause of allergic contact dermatitis (Fisher, 1986).

A 2% solution of methylmercury thioacetamide was used for topical treatment of a fungus infection of the skin; three cases of methylmercury poisoning occurred with serious encephalopathy (Okinaka et al., 1964). Friberg et al. (1961) found that methylmercury dicyanamide and mercury chloride penetrated the skin of guinea pigs at about the same rate.

Mercury vapour may penetrate the skin. Exposure of the forearm of human volunteers to radioactive mercury vapour indicated that an air concentration of 1 mg m^{-3} would result in a cutaneous uptake of 0.01–0.04 ng cm^{-2} min^{-1}, for a whole-body exposure, this uptake corresponded to about 2% of the uptake by inhalation of the same mercury concentration, but about half of the cutaneous mercury was shed by desquamation of skin cells during the following weeks, thus considerably diminishing the amount that, at a slow rate, was actually systemically absorbed (Hursh et al., 1989).

Current occupational exposure conditions would hardly warrant a general 'skin' denotation for mercury and its compounds. The use of the most toxic compounds has been restricted, thus further limiting the risk.

Lead

Apart from tetraethyllead, few lead compounds have been studied. A lead acetate solution in water and lead naphthenate in ether:ethanol (1:3) were applied to the shaven backs of rats; two days after five 500-μl doses on alternate days, increased lead concentrations were reported in the tissues (Rastogi and Clausen, 1976). Although the results suggest that both compounds can penetrate the skin, this report does not allow any quantitative assessment. Some penetration of lead nitrate through the skin has recently been suggested (Florence *et al.*, 1988).

Arsenic

Some arsenic compounds are still used as pesticides. Cacocylic acid (dimethylarsinic acid) and monosodium methanearsonate (MSMA) are used to kill selected trees during thinning of conifer forests; arsenic excretion in urine in forestry workers was found to reflect working habits and degree of skin protection (Norris, 1985). The dermal and oral LD_{50} values for cacocylic acid are of the same order of magnitude, while calcium arsenate has a higher oral toxicity (RTECS, 1983). Within 72 hours after application of MSMA or disodium methanearsonate (DSMA) to the skin of adult rats, about 10% had penetrated the skin (Shah *et al.*, 1987). Thus, with pesticide applicators, a certain risk of percutaneous absorption of arsenic compounds would be expected. Such exposures need to be limited as much as possible, as arsenic and its compounds are considered carcinogenic (class 1) (IARC, 1987).

Decaborane

The dermal LD_{50} for decaborane in the rat is 740 mg kg^{-1} and much higher than the value seen with oral exposure, but in the rabbit the dermal LD_{50} is 71 mg kg^{-1} (Svirbely, 1955). Human intoxications are much more likely to occur as a result of inhalation (Lowe and Freeman, 1957).

Sodium fluoroacetate

Low dermal LD_{50} values for sodium fluoroacetate have been documented in rodents (RTECS, 1983). Although a skin hazard may exist in humans, no evidence is available. Human toxicity has been proven by poisoning cases due to ingestion (Harrison *et al.*, 1952).

References

ACGIH (1986), *Documentation for the threshold limit values and biological exposure indices*, 5th edn (Cincinnati: American Conference of Governmental Industrial Hygienists).

BALLANTYNE, B. (1983), The influence of exposure route and species on the acute lethal toxicity and tissue concentrations of cyanide. In *Developments in the science and practice of toxicology*, eds A. W. Hayes, R. S. Schnell and T. S. Miya, pp. 583-6 (Amsterdam: Elsevier).

BENTUR, Y., KOREN, G., MCGUIGAN, M. and SPIELBERG, S. P. (1988), An unusual skin exposure to copper, clinical and pharmacokinetic evaluation, *Clinical Toxicology*, **26**, 371-80.

BOURGEOIS, M., DOOMS-GOOSSENS, A., KNOCKAERST, D., SPRENGERS, D., VAN BOVEN, M. and VAN TITTELBOOM, T. (1986), Mercury intoxication after topical application of a metallic mercury ointment, *Dermatologica*, **172**, 48-51.

CARSON, B. L. and SMITH, I. C. (1977), *Thallium, an appraisal of environmental exposure*, (Technical report No. 5) (Kansas City, MO: Midwest Research Institute).

CORSI, G. G. and PICOTTI, G. (1965), Clinical analysis of case histories of tetraethyl lead poisoning, *Folia Medica (Plovdiv)*, **48**, 856.

DUGARD, P. H. and SCOTT, R. C. (1984), Absorption through the skin. In *The Chemotherapy of Psoriasis*, ed. H. P. Baden, (*International Encyclopedia of Pharmacology and Therapeutics*, Section 110), (Oxford: Pergamon).

FISHER, A. A. (1986), *Contact dermatitis*, 3rd edn (Philadelphia, Lea & Febiger).

FLORENCE, T. M., LILLEY, S. G. and STAUBER, J. L. (1988), Skin absorption of lead, *Lancet*, ii, 157-8.

FRIBERG, L., SKOG, E. and WAHLBERG, J. E. (1961), Resorption of mercuric chloride and methyl mercury dicyandiamide in guinea-pigs through normal skin and through skin pre-treated with acetone, alkylaryl-sulphonate and soap, *Acta Dermato-Venereologica*, **41**, 40-52.

GLÖMME, J. and SJÖSTRÖM, B. (1955), Occupational thallium poisoning (in Swedish), *Läkartidningen*, **52**, 1436-41.

GRAHAM, D. L., LAMAN, D., TEODORE, J. and ROBIN, E. D. (1977), Acute cyanide poisoning complicated by lactic acidosis and pulmonary edema, *Archives of Internal Medicine*, **137**, 1051-5.

GRANDJEAN, P. (1984), Organolead exposures and intoxications. In *Biological effects of organolead compounds*, ed. P. Grandjean, pp. 227-41 (Boca Raton: CRC Press).

HARRISON, J. W. E., AMBRUS, J. L., AMBRUS, C. M., REES, E. W., PETERS, R. H., REESE, L. C. and BAKER, T. (1952), Acute poisoning with sodium fluoroacetate (compound 1080), *Journal of the American Medical Association*, **149**, 1520-2.

HOLTZMAN, N. A., ELLIOT, D. A. and HELLER, R. H. (1966), Copper intoxication, report of a case with observation on ceruloplasmin, *New England Journal of Medicine*, **275**, 347-52.

HURSH, J. B., CLARKSON, T. W., MILES, E. F. and GOLDSMITH, L. A. (1989), Percutaneous absorption of mercury vapor by man, *Archives of Environmental Health*, **44**, 120-7.

IARC (1987), *Overall evaluations of carcinogenicity: An updating of IARC Monographs Volumes 1 to 42* (IARC Monographs on the evaluation of carcinogenic risk of chemicals to man (Suppl. 7.) (Lyon: International Agency for Research on Cancer).

IRPTC (1982), *Thallium* (Scientific reviews of Soviet literature on toxicity and hazards of chemicals, No. 17) (Moscow: Centre of international projects, GKNT).

KEHOE, R. A. (1927), The toxicity of tetraethyl lead and inorganic lead salts, *Journal of Laboratory and Clinical Medicine*, **12**, 554-60.

KEHOE, R. A. and THAMANN, F. (1931), The behaviour of lead in the animal organism, II. Tetraethyllead. *American Journal of Hygiene*, **13**, 478-98.

LAUG, E. P. and KUNZE, F. M. (1948), The penetration of lead through the skin, *Journal of Industrial Hygiene*, **30**, 256.

LOWE, H. J. and FREEMAN, G. (1957), Boron hydride (borane) intoxication in man, *Archives of Industrial Health*, **16**, 523-33.

NIOSH (1976), *Criteria for a recommended standard... Occupational exposure to hydrogen cyanide & cyanide salts* (NaCN, KCN & Ca(CN)$_2$), (DHEW (NIOSH) Publication No. 77-108) (Cincinnati: National Institute for Occupational Safety and Health).

NORRIS, L. A. (1985), Exposure of applicators to monosodium methanearsonate and cacodylic acid in forestry. In *Dermal exposure related to pesticide use*, eds R. C. Honeycutt, G. Zweig and N. N. Ragsdale, (ACS symposium series 273), pp. 109–21 (Washington, DC: American Chemical Society).

OKINAKA, S., YOSHIKAWA, M., MOZAI, T., MIZUNO, Y., TERAO, T., WATANABE, H., OGIHARA, K., HIRAI, S., YOSHINO, Y., INOSE, T., ANZAI, S. and TSUDA, M. (1964), Encephalomyelopathy due to an organic mercury compound, *Neurology*, **14**, 69–76.

POTTER, A. L. (1950), The successful treatment of two recent cases of cyanide poisoning, *British Journal of Industrial Medicine*, 125–30.

PRICK, J. J. G., SMITT, W. G. S. and MULLER, L., eds (1955), *Thallium poisoning* (Amsterdam: Elsevier).

RASTOGI, S. C. and CLAUSEN, J. (1976), Absorption of lead through the skin, *Toxicology*, **6**, 371–6.

RICHESON, E. M. (1958), Industrial thallium poisoning, *Industrial Medicine and Surgery*, **27**, 607–19.

RTECS (1983), *Registry of toxic effects of chemical substances* (Cincinnati, OH: National Institute of Occupational Safety and Health).

SCHAEFER, H., ZESCH, A. and STÜTTGEN, G. (1982), *Skin permeability* (Berlin: Springer Verlag).

SCHAMBERG, J. F., KOLMER, J. A., RAIZISS, G. W. and GAVRON, J. L. (1918), Experimental studies of the mode of absorption of mercury when applied by inunction, *Journal of the American Medical Association*, **70**, 142–5.

SEAWRIGHT, A. A., BROWN, A. W., NG, J. C. and HRDLICKA, J. (1984), Experimental pathology of short-chain alkyllead compounds. In *Biological effects of organolead compounds*, ed. P. Grandjean, pp. 177–206 (Boca Raton: CRC Press).

SHAH, P. V., FISHER, H. L., SUMLER, M. R., MONROE, R. J., CHERNOFF, N. and HALL, L. L. (1987), Comparison of the penetration of 14 pesticides through the skin of young and adult rats, *Journal of Toxicology and Environmental Health*, **21**, 252–66.

SKOG, E. and WAHLBERG, J. E. (1964), A comparative investigation of the percutaneous absorption of metal compounds in the guinea pig by means of the radioactive isotopes ^{51}Cr, ^{58}Co, ^{65}Zn, ^{210m}Ag, ^{115m}Cd, ^{203}Hg, *Journal of Investigative Dermatology*, **43**, 187–92.

SVIRBELY, J. L. (1955), Toxicity tests of decaborane for laboratory animals, I. Acute toxicity studies, *Archives of Industrial Health*, **11**, 132–7.

TOVO, S. (1955), Poisoning due to KCN absorbed through the skin (in Italian), *Minerva Medico*, **75**, 158–61.

WALTON, D. C. and WITHERSPOON, M. G. (1926), Skin absorption of certain gases, *Journal of Pharmacology and Experimental Therapeutics (Baltimore)*, **26**, 315–24.

Chapter 3
Simple aliphatic compounds

Introduction

The simple aliphatic compounds most frequently considered to be skin penetrants are listed in table 3.1. The most important of these are described in the section below.

Among the alkanes, only n-hexane has received a 'skin' denotation, and only in one country.

The penetration of water-miscible alcohols increases as the lipid solubility increases (methanol constitutes an exception because it is highly caustic)

Table 3.1 Aliphatic hydrocarbons, alcohols, aldehydes, ketones, acids, ethers and esters which are considered a skin hazard in several countries (y = yes, n = no, o = other regulation). Those that are described in separate sections are underlined.

CAS No.	Chemical	Number of countries	FRG	Sweden	USA
67–56–1	*Methanol*	13	y	y	y
71–36–3	1-Butanol	8	o	y	y
108–11–2	Methyl isobutyl carbinol (4-methyl-2-pentanol)	9	y	n	y
107–18–6	*Allyl alcohol* (2-propen-1-ol)	15	y	y	y
107–19–7	Propargyl alcohol (2-propyn-1-ol)	10	y	n	y
109–86–4	*Ethylene glycol, monomethyl ether (2-methoxyethanol)*	9	y	y	y
110–80–5	Ethylene glycol, monoethyl ether (2-ethoxyethanol)	12	y	y	y
111–76–2	Ethylene glycol, monobutyl ether (2-butoxyethanol)	10	y	y	y
13429–07–7	Dipropylene glycol, monomethyl ether (1-(2-methoxypropoxy)-2-propanol)	7	o	n	y
111–15–9	Ethylene glycol, monoethyl ether, acetate (2-ethoxyethyl acetate)	10	y	y	y
110–49–6	Ethylene glycol, monomethyl ether, acetate (2-methoxyethyl acetate)	12	y	y	y
96–33–3	Methyl acrylate	12	o	y	y
140–88–5	Ethyl acrylate	10	y	y	y

47

(Scheuplein and Bronaugh, 1983). In contrast, addition of another polar group reduces the penetration rate by about 50-fold. Also, the rate decreases if a functional group is substituted by a more polar one. When increasing the length of an aliphatic alcohol, the membrane solubility increases, and thus also the permeability. More than 10 methylene groups in the carbon chain of the alcohol adds little more to the solubility and ultimately reduces the permeability due to the lower mobility of larger molecules. When pure alcohols are applied on the skin, the penetration rates are much lower than the corresponding rates for aqueous solutions. Esters penetrate faster than do the corresponding alcohols, and the unsubstituted hydrocarbons even faster.

Addition alcohols with a 'skin' denotation in some countries include propyl alcohol (1-propanol), and isopropyl alcohol (2-propanol). The latter may cause allergic contact dermatitis (Fisher, 1986). Monoalcohols from 1-butanol to 1-undecanol are skin irritants in concentrations above 50% (Jacobs *et al.*, 1987). Ethylene glycol (1,2-ethanediol) is also considered a skin hazard in some countries; the low vapour pressure of this solvent would perhaps in some situations promote the dermal uptake.

Aldehydes and ketones listed in a few countries are crotonaldehyde (2-butenal), methyl butyl ketone (2-hexanone), and methyl isobutyl ketone (4-methyl-2-pentanone).

With glycol ethers and their derivatives, the penetration rate for human skin *in vitro* decreased with increasing molecular size; diethylene glycol ethers penetrated the skin more slowly than their corresponding monoethylene glycol ether equivalents (Dugard *et al.*, 1984). Ethylene glycol ethers and their acetates are generally non-irritating, but ethylene glycol, monomethyl ether may act synergistically with weak irritants in solvent mixtures (Jacobs *et al.*, 1987). An additional glycol ether derivative listed as a skin hazard is ethylene glycol, monobutyl ether, acetate (2-butoxyethyl acetate). Dibutyl ether is the only other ether listed, and only in one country.

Due to caustic effects, formic acid, and peracetic acid are considered to be a skin hazard, and the same applies to several esters and peroxides: *tert*-butyl peracetate ((2-methyl-2-propyl)peracetate), *tert*-butyl hydroperoxide, di-*tert*-butyl peroxide (di(2-methyl-2-propyl)peroxide), dilauroyl peroxide, and methyl ethyl ketone peroxide (2-butanone peroxide).

Among the esters, methyl acetate, ethyl acetate, and ethyl acetoacetate may cause irritant contact dermatitis on prolonged dermal exposure, and this effect may be the basis for their inclusion among cutaneous hazards.

Percutaneous absorption of methyl methacrylate has been suggested (Rajaniemi *et al.*, 1989), and local neurotoxicity has been produced as a result of dermal exposure (Verkkala *et al.*, 1983). Both methyl methacrylate, and (2-hydroxyethyl)acrylate are included in some lists, primarily as important sensitizers (Fisher, 1986); some patients may be sensitive to several acrylates at the same time (Kanerva *et al.*, 1988). Turpentine oil is also an important sensitizer (Fisher, 1986). 1,3-Butadiene has probably been listed as a skin hazard because it is a class 2B carcinogen (IARC, 1987).

Aliphatic hydrocarbons

Because of the differences in toxicity, n-hexane is the only aliphatic hydrocarbon that is sometimes regarded as a skin absorption hazard. The solubility of this compound in psoriasis scales suggests a slow dermal uptake (Hansen and Andersen, 1988). However, on the basis of physicochemical properties alone, a penetration rate of $0.31\,\text{mg cm}^2\text{h}^{-1}$ was calculated; these calculations also suggested a possible significance of percutaneous absorption of the vapours (Fiserova-Bergerova and Pierce, 1989). Percutaneous penetration of n-hexane under occlusion was demonstrated in guinea pigs; this solvent resulted in lower concentrations in blood (in $\mu\text{g ml}^{-1}$) than several other common solvents (Jakobsen *et al.*, 1982). In rats, hexane penetrated more rapidly than cyclohexane and several aromatic compounds (Valette and Cavier, 1954). However, the *in vitro* penetration rate through excised rat skin was much lower (about $60\,\mu\text{g cm}^{-2}\text{h}^{-1}$) than for benzene and other aromatic compounds (Tsuruta, 1982). With this apparent discrepancy between the results, no definite conclusion can be made at this time.

Neurotoxic effects of n-nonane have been suggested. Thus, white spirit products containing this compound resulted in local neurotoxicity as a result of dermal exposure of the tail in rats (Verkkala *et al.*, 1984). The *in vitro* penetration rate for n-pentane through excised abdominal rat skin was $2.2\,\mu\text{g cm}^{-2}\text{h}^{-1}$ and even lower for 2-methylpentane, n-heptane and n-octane (Tsuruta, 1982).

Methanol

Skin exposure potential

Methanol is primarily used as raw material for the production of several organic compounds (mainly formaldehyde), but is also used widely as a solvent in paints and varnishes and for laboratory procedures. Methanol is finding increasing use as an additive to petrol. The total production within the EC was 2.5 million tonnes in 1983, i.e. about two-thirds of the actual consumption (Environmental Chemicals Data Information Network, ECDIN, Pers. comm., 1987). In the USA, 175 000 workers are potentially exposed to methanol (NIOSH, 1976).

Physicochemical properties

Methanol is a liquid at room temperature with a vapour pressure of 92 mm Hg. Methanol is miscible with water and most organic solvents. The octanol/water partition coefficient ($\log P_{ow}$) had been reported as -0.66 and -0.82, thus indicating that the solvent is primarily hydrophilic. Measured solubility in psoriasis scales (Hansen and Andersen, 1988) suggests that skin uptake will not be rapid. Assuming no toxic effect of methanol on the stratum corneum, a penetration rate of $2\,\text{mg cm}^{-2}\text{h}^{-1}$ was calculated by Fiserova-Bergerova and Pierce (1989) on the basis of physicochemical properties. However, delipidization

delipidization of the skin would tend to increase the absorption rate while evaporation would tend to limit the absorption from unoccluded application.

Experimental data

The skin penetration of methanol has been studied in detail *in vitro* (Treherne, 1956). Thus, the penetration of rabbit skin was only delayed by a few minutes, but the rate was about 50-fold below the rate seen for dermis alone (the outer layers removed). Thus, not surprisingly, the skin barrier against methanol absorption depends on the undamaged skin. The relative rate of skin penetration was independent of the methanol concentration in water (up to $12 \, mol \, l^{-1}$). The permeability constant was $1.96 \times 10^3 \, cm \, min^{-1}$ (Treherne, 1956). More recently, a lower average permeability constant of $2.6 \times 10^3 \, cm \, h^{-1}$ ($27 \, cm \, min^{-1}$) was found for human abdominal stratum corneum; the results varied widely with a coefficient of variation of 71% for eight specimens (Southwell *et al.*, 1984). Other *in vitro* studies have indicated that pure methanol may penetrate the skin faster than do aqueous solutions (Scheuplein and Blank, 1971). The high absorption rate may be due to damage of the skin caused by methanol, and this effect also facilitates subsequent penetration by other compounds. The absorption rate for pure methanol through human skin *in vitro* was about $8 \, mg \, cm^{-2} \, h^{-1}$ for epidermis alone and about twice as fast for a sample of full-thickness skin tissue (Scheuplein and Blank, 1973). McCord (1931) reported that all animals exposed to methanol by skin exposure under occlusion had died, the lowest lethal dose being $0.4 \, ml \, kg^{-1}$ for one monkey (partial occlusion only); dilated pupils occurred within 2 h. The report also stated that certain breeds of rabbits were less susceptible than were monkeys and rats. Unfortunately, confirmatory evidence is unavailable. NIOSH (1976) cited a dermal LD_{50} value of $20 \, g \, kg^{-1}$ in rabbits, a level several-fold above those seen after oral or intraperitoneal exposures. Systemic toxicity, including development of acidosis and neurotoxicity, and associated mechanisms have been studied in detail in animal models.

Human data

An experimental study suggested an average absorption rate of about $0.2 \, mg \, cm^{-2} \, min^{-1}$ ($12 \, mg \, cm^{-2} \, h^{-1}$) through the forearm skin of human volunteers (Dutkiewicz *et al.*, 1980). Accordingly, immersion of one hand in methanol for 2 min could lead to an absorption similar to the respiratory uptake from inhalation of $50 \, mg \, m^{-3}$ for 8 h (Dutkiewicz *et al.*, 1980). After dermal application of 15 ml of methanol under occlusion, high urinary methanol excretion levels were observed, corresponding to a 4-h inhalation exposure to 200 p.p.m. in air (Nakaaki *et al.*, 1980). Several fatal cases of methanol poisoning have been documented in children treated with methanol on the skin with occlusion (Gimenez *et al.*, 1968; Kahn and Blum, 1979). Respiratory or oral exposures in these cases were negligible. A painter who wore clothes and shoes soaked with methanol became blind a few days later (Dutkiewicz *et al.*, 1980).

Excessive urinary excretion of methanol has been documented in human volunteers after cutaneous exposure; the increase was detected 2 h after the initiation of exposure and reached a maximum at 6 h (Tada *et al.*, 1975). In occupational exposures, skin absorption is frequently not accounted for, but methanol has been shown to cause systemic toxicity, mainly affecting the central nervous system with headache, blurred vision, and nausea progressing to CNS depression and subsequent retinal damage with possible blindness (NIOSH, 1976). In addition, skin exposure to methanol delipidizes the skin and may cause severe dermatitis (NIOSH, 1976).

Conclusion

Methanol may rapidly penetrate the skin and may also cause severe local damage to the skin, thereby decreasing the effectiveness of the skin barrier. Due to the volatility of methanol, severe intoxications have only been documented in cases where the exposed skin was occluded. However, because of the rapid penetration and the systemic toxicity, methanol should be regarded a skin exposure hazard.

Allyl alcohol

Skin exposure potential

Allyl alcohol (2-propen-1-ol) and several derivatives are used as constituents of pharmaceuticals, perfumes and food flavourings. However, the largest single use of allyl alcohol is in the production of allyl esters as monomers and prepolymers. Within the EC, most production appears to take place in France, where 2000 tonnes were manufactured in 1973. Although published data are very limited, a skin exposure potential is probably present in the various manufacturing and user industries.

Physicochemical properties

Allyl alcohol is a colourless liquid at room temperature and has a vapour pressure of 24 mm Hg at 25°C. It has a pungent mustard-like odour. The boiling point is 97°C. This alcohol is miscible with water, ethanol and ether. Calculated solubility in psoriasis scales (Hansen and Andersen, 1988) suggests that skin penetration will be more rapid than with several other alcohols. Percutaneous absorption would be favoured by the rather small molecular size (molecular weight of 58) and the hydrophilic and lipophilic properties of this compound.

Experimental data

Allyl alcohol has been administered as a dermal dose to rabbits; skin penetration may lead to serious systemic injury, including visceral congestion periportal congestion of the liver, haematuria and nephritis (ACGIH, 1986). The LD_{50}

values for oral administration varied between 52 and 71 mg kg^{-1}, and for dermal administration between 53 and 89 mg kg^{-1} (Smyth and Carpenter, 1948; Spector, 1956; Dunlap *et al.*, 1958). Thus, allyl alcohol appears to penetrate the skin very efficiently. Allyl alcohol is a strong skin irritant, and dermal absorption may cause burns and deep pain at the site of absorption in addition to systemic toxicity (ACGIH, 1986). Unfortunately, no experimental information is available concerning the kinetics of skin absorption of this alcohol. Several studies have documented the capability of allyl alcohol to induce periportal (and central) liver necrosis (Dunlap *et al.*, 1958). This effect may be caused by the actions of acrolein which is the primary oxidation product of allyl alcohol. Other investigations using experimental animals have documented nephrotoxic effects (Carpanini *et al.*, 1978).

Human data

The exposure limit for allyl alcohol vapour is rather low, usually 2 p.p.m. or less, due to the irritant properties of this substance. Respiratory exposure below the limit for airway irritation is unlikely to cause any systemic toxicity. Because of prevention against skin injury to allyl alcohol, no information is available on systemic effects of dermal exposures. Data on systemic toxicity in humans are limited.

Conclusion

Allyl alcohol efficiently penetrates the skin in experimental studies. Although data on humans are lacking, this alcohol would be expected to constitute a significant skin absorption hazard which could result in serious systemic toxicity.

Other alcohols

The water-miscible monohydroxy alcohols have similar permeability constants and most likely penetrate the stratum corneum by the same mechanism as water; propanol, butanol and pentanol penetrate human skin *in vitro* somewhat faster when applied in aqueous solution, and even faster when applied in olive oil (Scheuplein and Blank, 1971). The solubility of most alcohols in psoriasis scales indicates the potential of skin penetration, but none of them have a high solubility (Hansen and Andersen, 1988).

1-Butanol is absorbed through the skin of dogs *in vitro* at a rate of 0.15 mg cm^{-1} h^{-1} (DiVincenzo and Hamilton, 1979) and through human skin *in vitro* at 0.02 mg cm^{-2} h^{-1} (Scheuplein and Blank, 1971). A 6 h exposure to a skin area of 3.14 cm^2 in guinea pigs resulted in blood concentrations that increased during the first 4 h to a plateau of about 2.4 mg l^{-1} (20 μmol l^{-1}) (Boman, 1989). Prior stripping of the stratum corneum, abrasion by sandpaper or scratching with a needle resulted in much higher blood concentrations of 1-butanol in guinea pigs after dermal exposure; a similar tendency was obtained with sensitized skin, while no change in 1-butanol absorption could be seen in irritant contact dermatitis (Boman, 1989). 1-Butanol can cause acute central nervous system (CNS) toxicity in experimental animals.

Adding additional polar groups drastically reduces the penetration rate. Thus, 2,3-butanediol has a k_p of 0.005×10^4 cm h^{-1}, as compared to 0.3×10^4 cm h^{-1} for 1-butanol; the same type of difference is seen between ethanol and ethylene glycol (Scheuplein and Blank, 1971). Polyfunctional alcohols penetrate even more slowly (Scheuplein and Bronaugh, 1983).

Methyl isobutyl carbinol (4-methyl-2-pentanol) is a primary irritant, and published dermal LD_{50} values do not suggest major skin penetration. However, propargyl alcohol (2-propyn-1-ol) has a dermal LD_{50} in the rabbit of 16 mg kg^{-1} (Clayton and Clayton, 1982), thus suggesting a possible hazard of skin contact.

Ketones

The solubility of several ketones in psoriasis scales indicates the potential of skin penetration, but none of them have a high solubility (Hansen and Andersen, 1988). However, these data are semiquantitative and do not allow predictions of cutaneous uptake rates.

A case of acetone intoxication has been described in a patient who had a lightweight cast set by means of a liquid mainly containing acetone; although inhalation of fumes probably contributed to the uptake, dermal absorption was also suggested by the latency period of about 9 h (Harris and Jackson, 1952). After dermal application of 15 ml of acetone under occlusion, acetone concentrations in blood were 4–12 μg ml^{-1}, corresponding to a 2-h inhalation exposure to 50–150 p.p.m. in air (Nakaaki et al., 1980).

Methyl ethyl ketone (butanone) has a calculated percutaneous absorption rate of 2.45 mg cm^{-2} h^{-1}, a rate that was estimated to cause a potentially significant absorption (Fiserova-Bergerova and Pierce, 1989). This solvent was absorbed much more easily through hydrated skin and more slowly through dehydrated skin; following continuous skin exposure to this solvent, methyl ethyl ketone could be detected in expired air of the subject in less than 3 min, but a steady-state was not reached until 2–3 h (Munies and Wurster, 1965). Skin penetration of this solvent is limited by high vapour pressure, and systemic toxicity is limited by rapid excretion.

Methyl n-butyl ketone (2-hexanone) is associated with the development of peripheral neuropathy; skin absorption of the radioactively labelled compound was demonstrated in two human volunteers at a rate of 4.8 and 8.0 μg min^{-1} cm^{-2}, the absorption rate being slightly lower for a 10% solution in methyl ethyl ketone (Di Vincenzo et al., 1978).

Ethylene glycol, monomethyl ether

Skin exposure potential

Ethylene glycol, monomethyl ether (2-methoxyethanol, methyl glycol) is a widely used solvent, but its utilization has recently decreased due to substitution by less toxic compounds, such as diethylene glycol, monomethyl ether. The main uses are as solvent in manufacture of protective coatings and also for printing inks, textile dyes, and leather finishes. Methyl glycol is used as anti-icing additive in brake fluids and as a fuel additive. In the USA, perhaps 100 000 workers are potentially exposed to methyl glycol (NIOSH, 1983).

Physicochemical properties

Methyl glycol is a colourless liquid at room temperature. It is miscible with both water and many organic solvents. With a boiling point of 124°C, the vapour pressure at room temperature is 6 mm Hg. Calculated solubility in psoriasis scales suggests that skin penetration could be rapid, but swelling of the scales in this solvent was less than predicted (Hansen and Andersen, 1988). The solubility properties as well as the small molecular size would favour skin penetration of this compound.

Experimental data

The dermal LD_{50} in rabbits is 1280 mg kg^{-1}, i.e. of the same order of magnitude as other administration routes (Anon., 1984). Although the acute toxicity is rather low, this solvent is more toxic than the longer-chain glycol ethers (Illing and Tinkler, 1985). *In vitro* studies using human abdominal skin have shown a steady-state penetration rate of 2.82 mg cm^{-2} h^{-1}, and this rate was higher than obtained with other glycol ethers; the penetration observed for methyl glycol is also faster than that seen with ethanol, but not as fast as methanol (Dugard *et al.*, 1984). The permeability constant (k_p) was 28.9 × 10^4 cm h^{-1}, i.e. higher than for water and for other glycol ethers tested (Dugard *et al.*, 1984). Methyl glycol appears to be non-irritant to skin of rabbits (Jacobs and Martens, 1985), but a slight decrease in the barrier function of human skin can be demonstrated as a result of prolonged methyl glycol exposure *in vitro* (Dugard *et al.*, 1984). While the acute toxicity is low in experimental animals, repeated percutaneous methyl glycol exposure has caused anaemia and damage to liver and kidneys (Hobson *et al.*, 1986). Reproductive toxicity has recently been reported as a result of skin exposure to methyl glycol (Wickramaratne, 1986; Hobson *et al.*, 1986).

Human data

After dermal application of 15 ml of methyl glycol under occlusion, rapid percutaneous absorption was observed, with blood levels reaching about 200 µg ml^{-1} after 2 h of skin exposure; this observation suggested a faster absorption than seen with methanol, acetone and methyl acetate (Nakaaki et al., 1980). Two workers in a textile printing plant developed bone marrow toxicity and toxic encephalopathy as a result of cutaneous exposures to methyl glycol in a plant where air concentrations were considered acceptable (Ohi and Wegman, 1978). Similarly, in other published cases, skin exposure contributed significantly to the development of methyl glycol toxicity (Zavon, 1963; Cohen, 1984). Skin irritation may occur following prolonged skin contact (Nakaaki et al., 1980).

Conclusion

Methyl glycol may rapidly penetrate the skin, and several human intoxications have been related to prolonged skin contact with this solvent. Methyl glycol should therefore be regarded a skin absorption hazard.

Other glycol derivatives

Glycol ethers are often used as solvents for cellulose ester, dyes, resins, lacquers and other surface coatings. Several of the glycol derivatives have solubilities in psoriasis scales that suggest the possibility of rapid skin penetration, although these data are only semi-quantitative (Hansen and Andersen, 1988). The acute toxicity of these compounds is low. 2-Methoxyethanol is more toxic than the longer chain glycols (Illing and Tinkler, 1985). Ethylene glycol, monomethyl ether; ethylene glycol, monomethyl ether, acetate; ethylene glycol, monoethyl ether; and ethylene glycol, monoethyl ether, acetate are teratogenic, embryotoxic and cause testicular damage (Johanson, 1988).

Glycol ethers penetrating through human skin *in vitro* can be listed in order of decreasing steady-state absorption: ethylene glycol, monomethyl ether > ethylene glycol, monopropyl ether > ethylene glycol, monoethyl ether = ethylene glycol, monoethyl ether, acetate > diethylene glycol, monomethyl ether > ethylene glycol, monobutyl ether > diethylene glycol, monoethyl ether > diethylene glycol, monobutyl ether (Dugard et al., 1984). The penetration rate decreased with increasing molecular size; diethylene glycol ethers penetrated the skin more slowly than their corresponding monoethylene glycol ether equivalents (Dugard et al., 1984).

Limited studies in beagle dogs confirm the skin penetration of ethylene glycol, monoethyl ether, acetate and ethylene glycol, monopropyl ether, acetate (Guest et al., 1984).

With ethylene glycol, monobutyl ether, immersion of two or four fingers for 2 h in the solvent resulted in a percutaneous uptake at a rate of 7–96 nmol cm^{-2} min^{-1} (50–680 µg cm^{-2} h^{-1}) without any apparent lag time (Johanson et al., 1988), i.e. similar to the rate observed *in vitro* where, however, a lag time was observed (Dugard et al., 1984). Immersion of four fingers in this solvent causes an absorption similar to the uptake resulting from inhalation of 20 p.p.m. during light work (Johanson et al., 1988). Dermal uptake in guinea pigs was enhanced when the glycol ether was present as an aqueous solution (Johanson and Fernström, 1988).

Assuming a continuous cutaneous exposure to undiluted solvent on 50cm^2 of skin, the percutaneous uptake would be larger than pulmonary uptake of vapours at the current exposure limit for: ethylene glycol, monomethyl ether; ethylene glycol, monoethyl ether; and ethylene glycol, monoethyl ether, acetate (Johanson, 1988). Thus, the two latter compounds would also appear likely to cause a skin absorption hazard.

References

ACGIH (1986), *Documentation for the threshold limit values and biological exposure indices.* 5th edn (Cincinnati: American Conference of Governmental Industrial Hygienists).

ANON (1984), Ethylene glycol monomethyl ether. *Dangerous Properties of Industrial Materials Report*, **4**, 67-70.

BOMAN, A. (1989), *Factors influencing the percutaneous absorption of organic solvents*, (Arbete och Hälsa Vol 1989: 11) (Stockholm: National Institute of Occupational Health).

CARPANINI, F. M. B., GAUNT, I. F., HARDY, J., GANGOLLI, S. D., BUTTERWORTH, K. R. and LLOYD, A. G. (1978), Short-term toxicity of allyl alcohol in rats, *Toxicology*, **9**, 29-45.

CLAYTON, G. D. and CLAYTON, F. E. (eds) (1982), *Patty's industrial hygiene and toxicology*, 3rd edn, Vol. 2C, p. 4672 (New York: Wiley).

COHEN, R. (1984), Reversible subacute ethylene glycol monomethyl ether toxicity associated with microfilm production, a case report, *American Journal of Industrial Medicine*, **6**, 441-6.

DIVINCENZO, G. D. and HAMILTON, M. L. (1979), Fate of n-butanol in rats after oral administration and its uptake in dogs after inhalation and skin application, *Toxicology and Applied Pharmacology*, **48**, 317-25.

DIVINCENZO, G. D., HAMILTON, M. L., KAPLAN, C. J., KRASAVAGE, W. J. and O'DONOGHUE, J. L. (1978), Studies on the respiratory uptake and excretion and the skin absorption of metyl n-butyl ketone in humans and dogs, *Toxicology and Applied Pharmacology*, **44**, 593-604.

DUGARD, P. H., WALKER, M., MAWDSLEY, S. J. and SCOTT, R. C. (1984), Absorption of some glycol ethers through human skin in vitro, *Environmental Health Perspectives*, **57**, 193-7.

DUNLAP, M. K., KODAMA, J. K., WELLINGTON, J. S., ANDERSEN, H. H. and HINE, C. H. (1958), The toxicity of allyl alcohol I. Acute and chronic toxicity, *Archives of Industrial Health*, **18**, 303-311.

DUTKIEWICZ, B., KONEZALIK, J. and KARWACKI, W. (1980), Skin absorption and per os administration of methanol in men, *International Archives of Occupational and Environmental Health*, **47**, 81-8.

FISEROVA-BERGEROVA, V. and PIERCE, J. T. (1989), Biological monitoring V. Dermal absorption, *Applied Industrial Hygiene*, **4**, F14-F21.

FISHER, A. A. (1986), *Contact dermatitis*, 3rd edn (Philadelphia, Lea & Febiger).

GIMENEZ, E. R., VALLEJO, N. E., ROY, E., LIS, M., IZURIETA, E. M., ROSSI, S. and CAPUCCIO, M. (1968), Percutaneous alcohol intoxication, *Clinical Toxicology*, **1**, 39-48.

GUEST, D., HAMILTON, M. L., DEISINGER, P. J. and DIVINCENZO, G. D. (1984), Pulmonary and percutaneous absorption of 2-propoxyethyl acetate and 2-ethoxyethyl acetate in beagle dogs, *Environmental Health Perspectives*, **57**, 177-83.

HANSEN, C. M. and ANDERSEN, B. H. (1988), The affinities of organic solvents in biological systems, *American Industrial Hygiene Association Journal*, **49**, 301-8.

HARRIS, L. C. and JACKSON, R. H. (1952), Acute acetone poisoning caused by setting fluid for immobilizing casts, *British Medical Journal*, **2**, 1024-6.

HOBSON, D. W., D'ADDARIO, A. P., BRUNER, R. H. and UDDIN, D. E. (1986), A subchronic dermal exposure study of diethylene glycol monomethyl ether and ethylene glycol monomethyl ether in the male guinea pig, *Fundamental and Applied Toxicology*, **6**, 339-48.

IARC (1987), *Overall evaluations of carcinogenicity: An updating of IARC Monographs Volumes 1 to 42*, (IARC Monographs on the evaluation of carcinogenic risk of chemicals to man (Suppl. 7.) (Lyon: International Agency for Research on Cancer).

ILLING, H. P. A. and TINKLER, J. J. B. (1985), Glycol ethers (Toxicity Review 10, Health and Safety Executive.) (London: Her Majesty's Stationery Office).

JACOBS, G. and MARTENS, M. (1985), *Skin irritation of aliphatic hydrocarbons, monocyclic aromatic hydrocarbons, alcohols, ethylene glycol ethers and their acetates* (Bruxelles: Institut d'Hygiène et d'Epidémiologie).

JACOBS, G., MARTENS, M. and MOSSELMANS, G. (1987), Proposal of limit concentrations for skin irritation within the context of a new EEC directive on the classification and labelling of preparations, *Regulatory Pharmacology and Toxicology*, **7**, 370-8.

JAKOBSON, I., WAHLBERG, J. E., HOLMBERG, B. and JOHANSSON, G. (1982), Uptake via the blood and elimination of 10 organic solvents following epicutaneous exposure of anesthetized guinea pigs, *Toxicology and Applied Pharmacology*, **63**, 181-7.

JOHANSON, G. (1988), Aspects of biological monitoring of exposure to glycol ethers, *Toxicology Letters*, **43**, 5-21.

JOHANSON, G., BOMAN, A. and DYNÉSIUS, B. (1988), Percutaneous absorption of 2-butoxyethanol in man, *Scandinavian Journal of Work and Environmental Health*, **14**, 101-9.

JOHANSON, G. and FERNSTRÖM, P. (1988), Influence of water on the percutaneous absorption of 2-butoxyethanol in guinea pigs, *Scandinavian Journal of Work and Environmental Health*, **14**, 95-100.

KAHN, A. and BLUM, D. (1979), Methyl alcohol poisoning in an 8-month-old boy: An unusual route of intoxication, *Journal of Pediatrics*, **94**, 841-3.

KANERVA, L. ESTLANDER, T. and JOLAKNI, R. (1988), Sensitization to patch test acrylates, *Contact Dermatology*, **18**, 10-5.

McCORD, C. P. (1931), Toxicity of methyl alcohol (methanol) following skin absorption and inhalation, a progress report, *Industrial Engineering Chemistry*, **23**, 931-6.

MUNIES, R. and WURSTER, D. E. (1965), Investigation of some factors influencing percutaneous absorption III. Absorption of methyl ethyl ketone, *Journal of Pharmaceutical Science*, **54**, 1281-4.

NAKAAKI, K., FUKABORI, S. and TADA, O. (1980), An experimental study on percutaneous absorption of some organic solvents, *The Journal of Science of Labour*, **56**, 1-9.

NIOSH (1976), *Criteria for a recommended standard... Occupational exposure to methanol*. (DHEW publication No. (NIOSH) 76-148) (Cincinnati, OH: National Institute of Occupational Safety and Health).

NIOSH (1983) *Glycol ethers, 2-methoxyethanol and 2-ethoxyethanol* (Current Intelligence Bulletin 39) (Cincinnati, OH: National Institute of Occupational Safety and Health).

OHI, G. and WEGMAN, D. H. (1978), Transcutaneous ethylene glycol monomethyl ether poisoning in the work setting, *Journal of Occupational Medicine*, **20**, 675-6.

RAJANIEMI, R., PFÄFFLI, P. and SAVOLAINEN, H. (1989), Percutaneous absorption of methyl methacrylate by dental technicians, *British Journal of Industrial Medicine*, **46**, 356-7.

SCHEUPLEIN, R. J. and BLANK, I. H. (1971), Permeability of the skin, *Physiology Review*, **51**, 702-47.

SCHEUPLEIN, R. J. and BLANK, I. H. (1973), Mechanism of percutaneous absorption IV. Penetration of nonelectrolytes (alcohols) from aqueous solutions and from pure liquids, *Journal of Investigative Dermatology*, **60**, 286-96.

SCHEUPLEIN, R. J. and BRONAUGH, R. L. (1983), Percutaneous absorption. In *Biochemistry and Physiology of the Skin*. ed. L. A. Goldsmith pp. 1255-95 (Oxford: Oxford University Press).

SMYTH, H. F., JR., and CARPENTER, C. P. (1948), Further experience with the range finding test in the industrial toxicology laboratory, *Journal of Industrial Hygiene and Toxicology*, **30**, 63-68.

SOUTHWELL, D., BARRY, B. W., WOODFORD, R. (1984), Variations in permeability of human skin within and between specimens, *International Journal of Pharmacology*, **18**, 299-309.

SPECTOR, W. S., ed. (1956), *Handbook of toxicology*, Vol. 1, Acute toxicities of solids, liquids and gases to laboratory animals, p. 14 (Philadelphia, PA).

TADA, O., NAKAAKI, K., FUKABORI, S. and YONEMOTO, J. (1975), An experimental study on the cutaneous absorption of methanol in man (in Japanese), *The Journal of Science of Labour (Jpn)*, **51**, 143-53.

TREHERNE, J. E. (1956), The permeability of skin to some non-electrolytes. *Journal of Physiology*, **133**, 171-80.

TSURATA, H. (1982), Percutaneous absorption of organic solvents III. On the penetration of hydrophobic solvents through the excised rat skin, *Industrial Health*, **20**, 335-45.

VERKKALA, E., RAJANIEMI, R. and SAVOLAINEN, H. (1983), Local neurotoxicity of methylmethacrylate monomer, *Toxicology Letters*, **18**, 111-4.

VERKKALA, E. PFAFFLI, P. and SAVOLAINEN, H. (1984), Comparison of local neurotoxicity of three white spirit formulations by cutaneous exposure of rat tail nerve, *Toxicology Letters*, **21**, 293-9.

WICKRAMARATNE, G. A. DE S. (1986), The teratogenic potential and dose-response of dermally administered ethylene glycol monomethyl ether (EGME) estimated in rates with the Chernoff-Kavlock assay, *Journal of Applied Toxicology*, **6**, 165-6.

ZAVON, M. R. (1963), Methyl cellosolve intoxication, *American Industrial Hygiene Association Journal*, **24**, 36-41.

Chapter 4
Halogenated aliphatic compounds

Introduction

The compounds considered are primarily the halogenated alkanes and alkenes. Those that are most frequently considered to be a skin hazard are indicated in table 4.1. The most important compounds are described in the sections below.

The majority of these compounds are used as solvents, but some use in pesticides also occurs. Thus, skin contact occurs frequently. However, most of the solvents have a high vapour pressure at room temperature, and the main problem in occupational use is usually inhalation of the vapours. When extensive skin contact does occur, in particular under occlusion, percutaneous absorption can be quite substantial, because many are relatively small molecules that are mainly lipophilic.

Solubility in psoriasis scales for various halogenated solvents suggests that skin penetration may be rapid for metnylene chloride, bromoform, ethylene dichloride, ethylene dibromide, 1,1,2,2-tetrachloroethane, 1,1,2,2-tetrabromoethane, perchloroethylene, 2,2-dichlorodiethyl ether, and bis(2-chloroisopropyl) ether (Hansen and Andersen, 1988).

Among the several compounds listed as a skin hazard in one or more countries, few have caused documented cases of human intoxication (Kimbrough *et al.*,

Table 4.1 Halogenated aliphatic and cyclic compounds with 'skin' denotation in several countries (y = yes, n = no, o = other regulation). Those that are described in separate sections are italicized.

CAS No.	Chemical	Number of countries	FRG	Sweden	USA
Halogenated alkanes					
56-23-5	*Carbon tetrachloride* (tetrachloromethane)	15	y	y	y
74-83-9	*Methyl bromide* (bromomethane)	12	y	y	y
75-25-2	Bromoform	8	o	y	y
74-88-4	*Methyl iodide* (iodomethane)	9	o	y	y
107-07-3	2-Chloroethanol (ethylenechlorohydrin)	11	y	y	y
79-00-5	1,1,2-Trichloroethane	8	y	n	y
79-34-5	1,1,2,2-Tetrachloroethane	11	y	n	y
67-72-1	Hexachloroethane	9	o	n	y
106-93-4	Ethylene dibromide (1,2-dibromoethane)	10	y	o	y
111-44-4	Bis(2-chloroethyl) ether	11	y	o	y
126-99-8	Chloroprene (2-chloro-1,3-butadiene)	11	y	y	y

1989). However, experimental studies indicate that increasing liver toxicity is associated with the ease with which a halogen can be removed to produce a reactive metabolite; a higher number of halogens in the molecule and increasing atomic weight of the halogen are therefore important (Zimmerman, 1978). Organs other than the liver may be affected as well.

In addition to the compounds in table 4.1, other halogenated alkanes with a 'skin' denotation in one or two countries are methylene chloride (dichloromethane), chloroform (trichloromethane), 1,1-dichloroethane, ethylene dichloride (1,2-dichloroethane), and 1,2-dibromo-3-chloropropane. As class 2B carcinogens (IARC, 1987), exposures to these compounds should be limited as much as possible; skin contact can also cause irritative contact dermatitis. Monobrominated and monochlorinated hydrocarbons with less than 16 carbons, as well as polyhalogenated compounds, such as dichloromethane, trichloromethane, 1,1,1-trichloroethane and tetrachloromethane, were tested and found to cause skin irritation (Jacobs et al., 1987). Exposure to these solvents may delipidize the skin and promote the absorption of other toxic substances.

Halogenated alkenes with a 'skin' denotation in a few countries include perchloroethylene and 1,3-dichloropropene that are class 2B carcinogens (IARC, 1987). Due to structural analogy, hexachloro-1,3-butadiene may be included in this group; this compound also has 'skin' denotation in some countries. Also with a 'skin' denotation, vinyl chloride (chloroethylene, chloroethene), is a confirmed human carcinogen (IARC, 1987). Further, trichloroethylene (trichloroethene) has received 'skin' denotation in single countries. The same applies to some halogenated alcohols, i.e. *alpha*-dichlorohydrin (1,3-dichloro-2-propanol), *beta*-dichlorohydrin (2,3-dichloro-1-propanol), and 2,2,3,3-tetrafluoro-1-propanol. Hexafluoroacetone (hexafluoropropanone) is also considered a skin hazard in a single country. A factor of possible importance in this regard is the delipidizing effects of cutaneous exposures to some of these compounds and the possible provocation of irritative contact dermatitis.

Carbon tetrachloride

Skin exposure potential

Carbon tetrachloride (tetrachloromethane) may be produced by several chlorination methods, and purification then includes neutralization, drying and distillation. Approximately 500 000 tonnes of carbon tetrachloride is produced in EC countries per year. The main uses have been as a chemical intermediate, grain fumigant, solvent and degreaser. Due to its toxicity, its use is being restricted.

Physicochemical properties

Carbon tetrachloride is a clear, colourless, non-flammable liquid. The boiling point is 76.7°C, and the vapour pressure is 91 mm Hg at 20°C. Carbon tetrachloride is almost insoluble in water, but miscible with common organic solvents. The octanol/water partition coefficient (log P_{ow}) is 2.64. Solubility in psoriasis scales was found to be better than predicted from solubility characteristics (Hansen and Andersen, 1988).

Experimental data

A dermal LD_{50} in the rat has been reported to be of similar magnitude as those obtained by other exposure routes (NIOSH, 1984). Percutaneous penetration of carbon tetrachloride was demonstrated in guinea pigs; under occlusion this and other solvents resulted in very similar concentrations in blood (0.8–1.9 μg ml^{-1} at 30 min) (Jakobson *et al.*, 1982). In mice, dermal application under occlusion resulted in less percutaneous penetration of carbon tetrachloride than seen with other chlorinated aliphatic solvents with a higher degree of water solubility (Tsuruta, 1975). Acute carbon tetrachloride intoxication results in neurotoxic effects, but subsequent hepatotoxicity may develop as well as toxic effects in the kidneys. Sufficient evidence is available that tetrachloromethane is carcinogenic to animals (IARC, 1987).

Human data

Immersion of a thumb in carbon tetrachloride for 30 min caused peak alveolar air concentrations close to 1 p.p.m., and immersion of both hands would conceivably lead to an absorption within the toxic range; the rate of penetration of human skin is about the same as for 1,1,1,-trichloroethane (Stewart and Dodd, 1964). Due to the high toxicity of carbon tetrachloride, percutaneous absorption can be of significance. Repeated or prolonged skin contact may also cause irritant eczema. Carbon tetrachloride exposure can cause severe lesions in the liver and kidneys (New *et al.*, 1962; Stewart *et al.*, 1963). Inadequate evidence is available on the carcinogenic risk to humans (IARC, 1987).

Conclusions

Absorption of significant amounts of carbon tetrachloride through the skin has been documented in studies of experimental animals and humans. Due to the high systemic toxicity of this solvent, the relatively slow percutaneous absorption is sufficient cause for regarding carbon tetrachloride a skin absorption hazard.

Methyl bromide

Skin exposure potential

Methyl bromide (bromomethane) may be produced by reacting methanol with hydrobromic acid. Methyl bromide is mainly used as a fumigant, in particular as a soil fumigant in greenhouses or space fumigant in warehouses, etc., additional uses as solvent, degreaser and methylating agent are of lesser importance. Several producers in Europe are known; the total consumption within the EC countries in 1979 was 6000 tonnes.

Physicochemical properties

Methyl bromide is a gas at room temperature, the boiling point is 3.6°C. A vapour pressure of two atmospheres is reached at 23°C. Methyl bromide is slightly soluble in water (0.9 g l^{-1} at 20°C) and easily soluble in ethanol, diethyl ether, carbon disulfide, chloroform, benzene and carbon tetrachloride. The octanol/water partition coefficient (log P_{ow}) is 1.19.

Experimental data

No published information on dermal absorption in experimental animals has been identified. Penetration of human skin by methyl bromide has been demonstrated *in vitro* (Winkel and Verspyck-Mijnssen, 1971). Experimental studies suggest that neurotoxicity may be the critical effect (Alexeeff and Kilgore, 1983). Limited evidence, in particular relating to mutagenic effects, is available regarding carcinogenicity of methyl bromide to experimental animals (IARC, 1986).

Human data

Percutaneous absorption of methyl bromide causes systemic toxicity similar to that seen after respiratory exposure, i.e. muscle weakness and pain, loss of coordination, convulsions and coma (Schifferli, 1942; Jordi, 1953). In one case, extensive skin contact with methyl bromide caused convulsions, and tremor which persisted 6 months later; the absence of respiratory symptoms suggests that inhalation exposure was insignificant (Longley and Jones, 1965). High concentrations of methyl bromide vapour during fumigation caused increased bromide levels in the blood, despite adequate respiratory protection; percutaneous absorption was the most likely pathway, as erythema and blistering documented local effects of the vapour (Zwaveling *et al.*, 1987). Methyl bromide vapour tends to adsorb to clothing and skin (van den Oever *et al.*, 1982). Skin contact with liquid methyl bromide, e.g. due to wet clothing, boots or gloves, may cause superficial burns with severe vesication, often delayed by several hours (Butler *et al.*, 1945; Alexeeff and Kilgore, 1983). Damage to liver and kidneys, and severe neurotoxicity may be seen in serious cases of intoxication (Hine, 1969; Alexeeff and Kilgore, 1983), but the significance of skin absorption has not been addressed in detail. Inadequate evidence is available concerning the possible carcinogenicity to humans (IARC, 1986).

Conclusions

Skin absorption of methyl bromide may take place when evaporation of liquid methyl bromide on the skin is limited or when high concentrations of vapour are present. The toxic potential of methyl bromide would clearly suggest that this compound be regarded a skin absorption hazard.

Methyl iodide

Skin exposure potential

Methyl iodide (iodomethane) may be produced by two major processes; within the EC, manufacturers are located in France, the Netherlands, Spain and the United Kingdom. Methyl iodide is primarily used as a methylating agent in pharmaceutical and other chemical synthesis; it is also used in microscopy due to its high refractive index and as a catalyst, e.g. in the production of organolead compounds.

Physicochemical properties

Methyl iodide is a liquid with a boiling point of 42.4°C and a vapour pressure of 400 mm Hg at 25°C. Methyl iodide is soluble in water (14 g l^{-1} at 20°C) and very soluble in acetone, ethanol, benzene, diethyl ether and carbon tetrachloride. The octanol/water partition coefficient (log P_{ow}) is about 1.69. It is an alkylating agent which reacts with water. The reactivity of methyl iodide would suggest that toxic effects can occur within the skin.

Experimental data

Skin application causes mild skin irritation (Buckell, 1950). Systemic toxicity may be delayed and includes narcosis and damage to the liver and kidneys (Buckell, 1950; IARC, 1986). Limited evidence is available that methyl iodide is carcinogenic in experimental animals (IARC, 1986).

Human data

Application of 1 ml of methyl iodide under a gauze dressing for 0.5 h on the skin of a human volunteer produced an erythematous reaction with vesicles 19 h after application; application without occlusion failed to produce any dermal toxicity (Buckell, 1950). Methyl iodide is primarily a central nervous system depressant and causes drowsiness, ataxia, tremor and coma; several case reports document the toxicity of methyl iodide (Garland and Camps, 1945; Appel *et al.*, 1975), but none provides any data on possible percutaneous absorption. Rubber gloves may not protect against skin exposure which can then lead to burning and blistering (Skutilova, 1975). No information is available concerning cancer in methyl iodide exposed workers (IARC, 1986).

Conclusions

No quantitative information is available on the percutaneous absorption of methyl bromide. However, this compound may result in severe skin damage and possible systemic toxicity, when evaporation is limited due to occlusion. It should therefore be regarded a skin absorption hazard.

Other compounds

Percutaneous absorption of most halogenated aliphatic compounds is limited due to their evaporation from unoccluded skin. However, with a relatively low molecular weight, compounds of somewhat hydrophilic character may be able to penetrate the skin in significant amounts. Thus, from physicochemical data, skin penetration rates of methyl chloroform (1,1,1,-trichloroethane) and trichloroethylene were calculated to be 1.64 and 0.27 mg cm^{-2}h^{-1}, respectively, both of potential importance in comparison with respiratory uptake; a lower calculated penetration rate of 0.11 mg cm^{-2}h^{-1} for perchloroethylene was considered insignificant (Fiserova-Bergerova and Pierce, 1989). LD$_{50}$ values suggest that 1,1,2-trichloroethane and 1,1,2,2-tetrachloroethane are among the chlorinated solvents with the highest acute toxicity; the dermal LD$_{50}$ values for these compounds are 7- and 20-fold, respectively, above the values obtained by oral administration (Smyth *et al.*, 1969).

Percutaneous application of 0.1 ml of various solvents to the skin of guinea pigs resulted in death of all 20 animals exposed to either 2-chloroethanol or 1,1,2-trichloroethane (Wahlberg and Boman, 1979). Cutaneous exposure to these two compounds rarely occurs because of the recognized toxicity. When compared to 1,1,1-trichloroethane (Jakobson *et al.*, 1982), the more hydrophilic 1,1,2-trichloroethane caused blood concentrations that were 3-fold higher after 6 h of dermal application under occlusion in guinea pigs (Jakobson *et al.*, 1977). In addition, the latter compound is much more toxic than the former. Percutaneous penetration of 1,1,1-trichloeoethane, perchloroethylene and trichloroethylene under occlusion in guinea pigs resulted in very similar concentrations in blood (0.8–1.9 µg ml^{-1} at 30 min) (Jakobson *et al.*, 1982).

Similar blood concentrations for 1,1,1-trichloroethane were obtained by Boman (1989) after a 6-h exposure to a skin area of 3.14 cm^2 in guinea pigs, and a peak of about 10 µmol l^{-1} (1.3 µg ml^{-1}) was reached after about 30 min. Prior stripping of the stratum corneum, abrasion by sandpaper or scratching with a needle resulted in lower blood concentrations after dermal exposure, presumably because of removal of lipophilic structures of the stratum corneum (Boman, 1989).

In mice, dermal application on 2.92 cm^2 abdominal skin under occlusion resulted in a percutaneous penetration of 50-fold more methylene chloride (dichloromethane) than perchloroethylene on a molar basis; intermediate absorption rates were seen with ethylene dichloride (1,2-dichloroethane), chloroform (trichloromethane), 1,1,2-trichloroethane, 1,1,2,2-tetrachloroethane, carbon tetrachloride, and 1,1,1-trichloroethane, in descending order (Tsuruta, 1975). Thus, the penetration rate increased with the degree of water solubility. Based on some preliminary calculations; tetrachloromethane, 1,1,2,2-tetrachloroethane, 1,1,2-trichloroethane, chloroform, ethylene dichloride and methylene chloride were considered potentially toxic hazards if both hands were immersed in the solvent (Tsuruta, 1975).

Immersion of one hand for 30 min in 1,1,1-trichloroethane resulted in a peak alveolar concentration of 22 p.p.m. in human volunteers, a level associated with inhalation of solvent vapours much below the exposure limit; similar conclusions were reached for trichloroethylene, perchloroethylene and methylene chloride (Stewart and Dodd, 1964).

Similarly, after dermal application of 1,1,1-trichloroethane under occlusion on a small area of 12.5 cm^2 of skin for 2 h in human volunteers, concentrations of the solvent in exhaled air reached 3–5 p.p.m., corresponding to those seen after a 2-h inhalation exposure to 10–20 p.p.m. in air (Nakaaki et al., 1980).

Also, when one hand was soaked for 5 min in perchloroethylene and 1,1,1-trichloroethane, solvent concentrations in the blood were up to 9.0 and 4.0 μmol l^{-1} and (1.5 and 0.5 μg ml^{-1}) or 130- and 35-fold above the levels in blood obtained from the contralateral arm; these levels were as high as blood levels seen after inhalation of solvent vapours at about 50 p.p.m. in the air (Aitio et al., 1984).

Immersion of one hand in trichloroethylene for 30 min resulted in an uptake which was about one-third of the absorption resulting from inhalation of solvent vapour at 100 p.p.m. for 4 h, but trichloroethylene concentrations in exhaled air during the first 2 h after the exposure were higher in relation to the dermal exposure (Sato and Nakajima, 1978).

Sufficient evidence is available that perchloroethylene, dichloromethane and chloroform are carcinogenic to animals, but inadequate evidence is available on the risk to humans (IARC, 1987).

With 2,2'-dichlorodiethyl ether (bis(2-chloroethyl) ether), low dermal LD$_{50}$ values suggest a possible skin hazard, but this compound also is a primary irritant. Sufficient evidence is available that it causes cancer in experimental animals (IARC, 1987).

Hexachloroethane has high dermal LD$_{50}$ values, but experimental evidence suggests that it may be a carcinogen (IARC, 1987).

Chloroprene (2-chloro-1,3-butadiene) causes temporary hair loss upon dermal contact (Flesch and Goldstone, 1951). Inadequate evidence is available on the possible carcinogenicity (IARC, 1987).

Methylene bromide (dibromomethane) vapour (at 500 to 10000 p.p.m.) rapidly penetrated the skin of rats and resulted in readily detectable levels in the blood; the less lipophilic bromochloromethane penetrated less rapidly (McDougall et al., 1985).

Bromoform (tribomomethane) was in one study found to penetrate the skin and cause detectable exhalation of the vapour after 0.5 h, i.e. as fast as trichloroethylene and faster than several other halogenated compounds (Schwander, 1936).

Ethylene dibromide (1,2-dibromoethane) is a primary irritant and has a dermal LD_{50} of about $300\,mg\,kg^{-1}$ in rabbits, volatility decreases the skin absorption potential (Rowe et al., 1952). Ethylene dibromide poisoning has been described in humans (Olmstead, 1960). It is a class 2A carcinogen (IARC, 1987).

Chlorinated paraffins may be used in cutting oils; one product (Cerechlor S56L) penetrated human skin in vitro at a slow rate of 0.03–$0.05\,\mu g\,cm^{-2}\,h^{-1}$, while no detectable penetration could be determined with another (Cerechlor S52) (Scott, 1989).

Alachlor, a chlorinated acetanilide (2-chloro-2',6'-diethyl-N-(methoxymethyl)acetanilide), is used as a herbicide. This compound showed a high affinity for human stratum corneum; following a lag time of 1.2–1.8 h, peak penetration rates through human skin in vitro increased with increasing dilutions in water and reached a maximum of $9.6\,\mu g\,cm^{-2}\,h^{-1}$ (Bucks et al., 1989).

References

AITIO, A., PEKARI, K. and JÄRVISALO, J. (1984), Skin absorption as a source of error in biological monitoring, *Scandinavian Journal of Work and Environmental Health*, **10**, 317–320.
ALEXEEFF, G. V. and KILGORE, W. W. (1983), Methyl bromide, *Residue Reviews*, **88**, 101–53.
APPEL, G. B., GALEN, R., O'BRIEN, J. and SCHOENFELDT, R. (1975), Methyl iodide intoxication. A case report, *Annals of Internal Medicine*, **82**, 534–6.
BOMAN, A. (1989), *Factors influencing the percutaneous absorption of organic solvents*, (Arbete och Hälsa Vol 1989: 11) (Stochholm: National Institute of Occupational Health).
BUCKELL, M. (1950), The toxicity of methyl iodide, I. Preliminary survey, *British Journal of Industrial Medicine*, **7**, 122–4.
BUCKS, D. A. W., WESTER, R. C., MOBAYEN, M. M., YANG, D., MAIBACH, H. I. and COLEMAN, D. L. (1989), In vitro percutaneous absorption and stratum corneum binding of alachlor: Effect of formulation dilution with water, *Toxicology and Applied Pharmacology*, **100**, 417–23.
BUTLER, E. C. B., PERRY, K. M. A. and WILLIAMS, J. R. F. (1945), Methyl bromide burns, *British Journal of Industrial Medicine*, **2**, 30–1.
FISEROVA-BERGEROVA, V. and PIERCE, J. T. (1989), Biological monitoring V. Dermal absorption, *Applied Industrial Hygiene*, **4**, F14–F21.
FLESCH, P. and GOLDSTONE, S. B. (1951), Depilatory action of the intermediary polymers of chloroprene, *Science*, **113**, 126–7.
GARLAND, A. and CAMPS, F. E. (1945), Methyl iodine poisoning, *British Journal of Industrial Medicine*, **2**, 209–11.
HANSEN, C. M. and ANDERSEN, B. H. (1988), The affinities of organic solvents in biological systems, *American Industrial Hygiene Association Journal*, **49**, 301–8.
HINE, C. H. (1969), Methyl bromide poisoning, *Journal of Occupational Medicine*, **11**, 1–10.
IARC (1986), *Some halogenated hydrocarbons and pesticide exposures*. (IARC Monographs on the evaluation of carcinogenic risk of chemicals to man, Vol. 41) (Lyon: International Agency for Research on Cancer).
IARC (1987), *Overall evaluations of carcinogenicity: An updating of IARC Monographs Volumes 1 to 42* (IARC Monographs on the evaluation of carcinogenic risk of chemicals to man, Suppl. 7) (Lyon: International Agency for Research on Cancer).

JACOBS, G., MARTENS, M. and MOSSELMANS, G. (1987), Proposal of limit concentrations for skin irritation within the context of a new EEC directive on the classification and labelling of preparations, *Regulatory Pharmacology and Toxicology*, **7**, 370-8.

JAKOBSON, I., HOLMBERG, B. and WAHLBERG, J. E. (1977), Variations in the blood concentration of 1,1,2-trichloroethane by percutaneous absorption and the other routes of administration in the guinea pig, *Acta Pharmacology and Toxicology*, **41**, 497-506.

JAKOBSON, I., WAHLBERG, J. E., HOLMBERG, B. and JOHANSSON, G. (1982), Uptake via the blood and elimination of 10 organic solvents following epicutaneous exposure of anesthetized guinea pigs, *Toxicology and Applied Pharmacology*, **63**, 181-7.

JORDI, A. U. (1953), Absorption of methyl bromide through the intact skin, a report of one fatal and two non-fatal cases, *Aviation Medicine*, **24**, 536-9.

KIMBROUGH, R. D., MAHAFFEY, K. R., GRANDJEAN, P., SANDOE, S. -H. and RUTSTEIN, D. D. (1989), *Clinical effects of environmental chemicals* (New York: Hemisphere).

LONGLEY, E. O. and JONES, A. T. (1965), Methyl bromide poisoning in man, *Industrial Medicine and Surgery*, **34**, 499-502.

MCDOUGALL, J. N., JEPSON, G. W., CLEWELL, H. J., III. and ANDERSEN, M. E. (1985), Dermal absorption of dihalomethane vapors, *Toxicology and Applied Pharmacology*, **79**, 150-8.

NAKAAKI, K., FUKABORI, S. and TADA, O. (1980), An experimental study on percutaneous absorption of some organic solvents, *Journal of Science Labour*, **56**, 1-9.

NEWS, P. S., LUBASH, G. D., SCHERR, L. and RUBIN, A. L. (1962) Acute renal failure associated with carbon tetrachloride intoxication, *Journal of the American Medical Association*, **181**, 903-6.

NIOSH (1984), *Registry of toxic effects*, (DHSS (NIOSH) Publication No. 83-107-4) (Cincinnati, OH: National Institute for Occupational Safety and Health).

VAN DEN OEVER, R., ROOSELS, D. and LAHAYE, D. (1982), Actual hazard of methyl bromide fumigation in soil disinfection, *British Journal of Industrial Medicine*, **39**, 140-4.

Olmstead, E. V. (1960), Pathological changes in ethylene dibromide poisoning, *Archive of Industrial Health*, **21**, 525-9.

ROWE, V. K., SPENCER, H. C., MCCOLLISTER, D. D., HOLLINGSWORTH, R. L. and ADAMS, E. M. (1952), Toxicity of ethylene dibromide determined on experimental animals, *Archives of Industrial Hygiene and Occupational Medicine*, **6**, 158-73.

SANGSTER, B. (1987), Exposure of the skin to methyl bromide, a study of six cases occupationally exposed to high concentrations during fumigation, *Human Toxicology*, **6**, 491-5.

SATO, A. and NAKAJIMA, T. (1978), Differences following skin or inhalation exposure in the absorption and excretion kinetics of trichloroethylene and toluene, *British Journal of Industrial Medicine*, **35**, 43-9.

SCHIFFERLI, E. (1942), Intoxications par le bromure de méthyle, *Revue Medicale de la Suisse Romande*, **62**, 244-50.

SCOTT, R. C. (1989), In vitro absorption of some chlorinated paraffins through human skin. *Archives of Toxicology*, **63**, 425-6.

SCHWANDER, P. (1936), Percutaneous diffusion of halogenated hydrocarbon compounds (in German), *Archiv für Gewerbepathologie und Gewerbehygiene*, **7**, 109-16.

SKUTILOVA, J. (1975), Acute impairment due to methyl iodide (in Czech), *Pracovni Lekaistri*, **27**, 341-2.

SMYTH, H. F., CARPENTER, C. P., WEIL, C. S., POZZANI, U. C., STRIEGEL, J. A. and NYCUM, J. S. (1969), Range-finding toxicity data. List VII, *American Industrial Hygiene Association Journal*, 470-6.

STEWART, R. D., BOETTNER, E. A., SOUTHWORTH, R. R. and CERNY, J. C. (1963), Acute carbon tetrachloride intoxication. *Journal of the American Medical Association*, **183**, 994–7.

STEWART, R. D. and DODD, H. C. (1964), Absorption of carbon tetrachloride, trichloroethylene, tetrachloroethylene, methylene chloride, and 1,1,1-trichloroethane through human skin, *American Industrial Hygiene Association Journal*, **25**, 439–46.

TSURUTA, H. (1975), Percutaneous absorption of organic solvents. 1. Comparative study of the in vivo percutaneous absorption of chlorinated solvents in mice, *Industrial Health*, **13**, 227–36.

WAHLBERG, J. E. and BOMAN, A. (1979), Comparative percutaneous toxicity of ten industrial solvents in the guinea pig, *Scandinavian Journal of Work and Environmental Health*, **5**, 345–51.

WINKEL, P. and VERSPYCK-MIJNSSEN, G. A. W. (1971), The part played by skin penetration in the entry of substances into the body (in Dutch), *Tijdschrift voor Sociale Geneeskunde*, **49**, 328–30.

ZIMMERMAN, H. J. (1978), *Hepatotoxicity, The adverse effects of drugs and other chemicals on the liver* (New York: Appleton-Century-Crofts).

ZWAVELING, J. H., DE KORT, W. A. M., MEULENBELT, J., HEZEMANS-BOER, M., VAN VLOTEN, W. A. and SANGSTER, B. (1987), Exposure of the skin to methyl bromide, a study of six cases occupationally exposed to high concentrations during fumigation, *Human Toxicology*, **6**, 491–5.

Chapter 5
Aliphatic amides, nitriles, and amines

Introduction

The aliphatic amides, nitriles, and amines most frequently considered to be skin penetrants are listed in table 5.1. The most important of these are described in the sections below. Among the several compounds listed as a skin hazard in one or more countries, few have caused documented cases of human intoxication (Kimbrough *et al.*, 1989). These compounds vary much in their toxicity.

Nitriles with a single 'skin' denotation and not listed in the table are acetonitrile, acetone cyanohydrin and malononitrile. Methacrylonitrile is listed in several countries.

Additional amines with a 'skin' denotation in one or two countries are methylamine, ethylamine, butylamines (other than 1-butylamine), allylamine (2-propenylamine), diethylamine, triethylamine, and ethanolamine (2-aminoethanol). 2-(n-Butyl)aminoethanol has a 'skin' denotation in several countries.

Solubility in psoriasis scales would suggest that ethylenediamine and diethylenetriamine could penetrate the skin quite rapidly (Hansen and Andersen, 1988).

Table 5.1 Aliphatic amines, amides, and nitriles that are considered a skin hazard in several countries (y = yes, n = no, o = other regulation). Those that are described in separate sections are italicized.

CAS No.	Chemical	Number of countries	FRG	Sweden	USA
Amides					
68-12-2	*N,N-Dimethylformamide*	13	y	y	y
127-19-5	*N,N-Dimethylacetamide*	9	y	n	y
79-06-1	*Acrylamide*	13	y	y	y
Nitriles					
3333-52-6	Tetramethylsuccinonitrile	7	y	n	y
107-13-1	*Acrylonitrile*	14	y	y	y
Amines					
109-73-9	1-Butylamine	12	y	y	y
108-18-9	Di-isopropylamine (bis(2-propyl)amine)	8	o	n	y
100-37-8	N,N-Diethyl-2-aminoethanol	9	y	n	y
111-40-0	Diethylenetriamine ((2-aminoethyl)-1,2-ethanediamine)	8	n	y	y
62-75-9	*N*-Nitrosodimethylamine (dimethylnitrosamine)	6	o	o	y

N,N-*Dimethylformamide*

Skin exposure potential

N,N-Dimethylformamide (DMF) is produced from dimethylamine. The total production within the EC countries in 1983 was 60 000 tonnes with a decreasing tendency. DMF is a universal industrial solvent, e.g. for acrylic fibres, polyurethanes, paints and PVC, but it is also used as an intermediate for drug, pesticide and other organic synthesis. In the USA, an estimated 94 000 workers are exposed to dimethylformamide (IARC, 1989).

Physicochemical properties

Dimethylformamide is a colourless liquid with a boiling point of 153°C. The vapour pressure at 25°C is 3.7 mm Hg. Dimethylformamide dissolves many inorganic salts and acids, and acetylene and other gases. This compound is miscible with water and with most solvents, but has limited solubility in aliphatic hydrocarbons. The octanol/water partition coefficient (log P_{ow}) is -1.01. The calculated and measured solubility in psoriasis scales is excellent, thus suggesting that skin uptake could be considerable (Hansen and Andersen, 1988). On the basis of physicochemical properties, and assuming no interference with the function of the stratum corneum as a barrier, a penetration rate of 1.03 mg cm^{-2} h^{-1} was calculated by Fiserova-Bergerova and Pierce (1989).

Experimental data

The dermal toxicity in the rabbit and the rat is slightly below that seen after other administration routes (Kennedy, 1986). When rat tails were dipped in dimethylformamide at 40°C for 8 h, all of seven rats died within 4 days of exposure (Massman, 1956); assuming a constant penetration through the skin, the absorption rate can be estimated as at least 7 mg cm^{-2} h^{-1}. The approximate lethal dose in rabbits (3.4 g kg^{-1}) was lower than that in the rat (17 g kg^{-1}) (Stula and Krauss, 1977). In another study, the minimal lethal dose for dimethylformamide applied on the skin under occlusion was 4.2 g kg^{-1} in rats and 2.8 g kg^{-1} in rabbits (Czajkowska, 1981). These levels are slightly above the LD$_{50}$ levels seen with other administration routes (Kennedy, 1986). Daily dermal doses of 2 g kg^{-1} in rabbits caused mortality which began to occur after the fifth dose (Kennedy and Sherman, 1986). Slight skin irritation has been recorded; the systemic effects include liver toxicity and a depressant effect on the nervous system (Kennedy, 1986). Dimethylformamide promotes the absorption of polar compounds by increasing the skin permeability, perhaps by displacing water in the stratum corneum (Munro and Stoughton, 1965; Kennedy, 1986). Following dermal application to the mother, embryotoxicity has been demonstrated in rats, but only at doses causing maternal toxicity (Stula and Krauss, 1977). Inadequate evidence is available for the carcinogenicity of dimethylformamide to animals (IARC, 1989).

Human data

In vitro studies using excised human skin have shown that concentrated dimethylformamide rapidly penetrates the skin barrier, and the penetration rate increases with time; aqueous solutions penetrate much more slowly, the penetration constant being much lower and, at decreasing concentrations, depending to a much lesser degree on the time (Bortsevich, 1984). In an acrylic fibre factory, absence of skin protection resulted in acute symptoms of dimethylformamide poisoning and a threefold increase in the urinary excretion of *N*-methylformamide (Lauwerys *et al.*, 1980). The urinary excretion of this metabolite correlated with the airborne exposure in this factory only when complete protection against skin exposure was employed (Lauwerys *et al.*, 1980). Similar results were obtained in an experimental study (Maxfield *et al.*, 1975). Recent research has shown that the main urinary metabolite is *N*-hydroxy-*N*-methylformamide which is converted to *N*-methylformamide during the analysis (Scailteur and Lauwerys, 1987); past studies which only assessed *N*-methylformamide must therefore be evaluated with caution. Cutaneous exposure may result in slight skin irritation (Massman, 1956). One case of dimethylformamide-related contact dermatitis has recently been described (Camarasa, 1987). In an accident a worker was splashed with dimethylformamide over about 20% of his body surface; dermal irritation and hyperaemia developed, and after almost 3 days severe abdominal pain ensued and liver damage and porphobilinogenuria were detected (Potter, 1973). However, some inhalation of dimethylformamide may have contributed to the toxicity. Percutaneous absorption was probably involved in other cases of poisoning where protective gloves were not used, or the solvent was splashed onto the body (Paoletti *et al.*, 1982). The liver seems to be the major target organ for dimethylformamide poisoning, and an early symptom is alcohol intolerance (Scailteur and Lauwerys, 1987). Possible induction of testicular cancer has been suggested (Ducatman *et al.*, 1986), but only limited evidence is available for the carcinogenicity to humans (IARC, 1989).

Conclusions

Dimethylformamide can penetrate the skin in significant quantities. This solvent will also promote the absorption of several other compounds through the skin. Dimethylformamide should therefore be regarded a skin exposure hazard.

N,N-*Dimethylacetamide*

Skin exposure potential

N,N-Dimethylacetamide is produced from acetic anhydride and dimethylformamide. It is widely used as a solvent for various plastic materials and in many organic reactions.

Physicochemical properties

Dimethylacetamide is a liquid with a boiling point of 166°C and a vapour pressure of 1.3 mm Hg at 25°C. It is miscible with water and most organic solvents. The octanol/water partition coefficient (log P_{ow}) is -0.77. In agreement with the amphoteric properties, the calculated solubility of this compound in psoriasis scales suggests that considerable skin uptake could occur (Hansen and Andersen, 1988). Assuming no effect on the barrier function of the stratum corneum, a percutaneous penetration rate of 1.00 mg cm^{-2} h^{-1} was calculated by Fiserova-Bergerova and Pierce (1989) on the basis of physicochemical properties.

Experimental data

Percutaneous absorption has been demonstrated using ^{14}C-labelled dimethylacetamide (Kennedy, 1986). The dermal LD$_{50}$ is 2.24 ml kg^{-1} in the rabbit (Smyth et al., 1962), i.e. only slightly above that seen with other administration routes (Kennedy, 1986). All rabbits which received daily dermal applications of 2 mg kg^{-1} died within five doses (Kennedy and Sherman, 1986). In dogs, slight systemic toxicity was seen after daily dermal application of 0.1 ml kg^{-1} for 6 months, and mortality was observed after several days of dermal applications of 4 ml kg^{-1} (Horn, 1961). Slight to severe skin irritation has been recorded (Horn, 1961; Kennedy, 1986). Dimethylacetamide promotes the absorption of polar compounds by increasing the skin permeability, perhaps by displacing water in the stratum corneum (Munro and Stoughton, 1965; Kennedy, 1986). Following dermal application to the mother, embryotoxicity was demonstrated in rats at doses which did not affect maternal weight or cause any signs of toxicity (Stula and Krauss, 1977). The liver is the target organ (Kennedy, 1986).

Human data

An experimental study in human volunteers indicated that about 30% of cutaneously applied dimethylacetamide was absorbed, as compared to 70% if inhaled; absorption was determined by the urinary excretion of the metabolite monomethylacetamide (Maxfield et al., 1975). Workers exposed to dimethylacetamide showed signs of significant percutaneous uptake, as measurement of urinary excretion of monomethylacetamide showed little relationship to the airborne levels of dimethylacetamide (Borm et al., 1987). Calculations of percutaneous and respiratory absorption rates suggest that the former may be of potential significance (Fiserova-Bergerova and Pierce, 1989). Although no detailed information is available on the contribution by skin exposure under occupational exposures, dermal absorption has been considered so significant that no air concentration will provide protection if skin contact is permitted (ACGIH, 1986). Hepatotoxicity is the most important effect due to occupational exposures (Corsi, 1971).

Conclusions

Dimethylacetamide may penetrate the skin in significant quantities. This solvent will also promote the absorption of several other compounds through the skin. Dimethylacetamide should therefore be regarded a skin exposure hazard.

Acrylamide

Skin exposure potential

Acrylamide (2-propeneamide) is mainly used in the production of various polymers and co-polymers. The total production capacity within the EC was 42500 tonnes in 1984. Based on estimated consumption and exports of polyacrylamide, the use of acrylamine in 1982 in Western Europe amounted to about 18000 tonnes. Major uses were water treatment, mineral processing, and pulp and paper production. Polyacrylamide and other polymers may contain considerable quantities of leachable acrylamide monomer. Although occupational exposures mainly occur in the production of acrylamide and the polymers of acrylamide users of acrylamide-containing polymers may be also dermally exposed to small amounts of acrylamide.

Physicochemical properties

Acrylamide is an odourless, solid material at room temperature, usually in the form of a white powder. The melting point is 84.5°C. The vapour pressure at 20°C is 0.007 mm Hg. Acrylamide is very water soluble, and 1 l of water at 30°C may dissolve more than 2 kg of the compound. Some lipophilicity is indicated by the fact that 1 l of acetone may dissolve 631 g, and 1 l of benzene dissolves 3.5 g of acrylamide (WHO, 1989).

Experimental data

Application of acrylamide for 4 h on the skin of rats showed an approximate LD_{50} level of 400 mg kg^{-1} body weight (Novikova, 1979), i.e. only slightly more than the LD_{50} levels of 100–200 mg kg^{-1} related to oral or parenteral dosages (WHO, 1985). Chronic toxicity in rats was encountered when the daily dermal application exceeded 0.5 mg kg^{-1} body weight; on the basis of this threshold, a tentative safe exposure limit was estimated at 3.5 mg for exposure to acrylamide in humans (Novikova, 1979). Reactions with biological molecules would tend to decrease the transport rate, but as much as 50% of the acrylamide present in the blood 24 h after skin exposure appears to be in unbound form (Hashimoto and Ando, 1975). Experimental studies using radioactively labelled acrylamide have shown that this compound rapidly penetrates the skin, partly through the hair follicles (Hashimoto and Ando, 1975). The permeability coefficient has not been estimated. Two hours after application of acrylamide (100 mg kg^{-1}) on the shaved skin of Swiss mice, glutathione depletion could be detected as an indication of toxicity in both skin and liver (Mukhtar et al., 1981). A 50% mortality was seen in mice exposed to a 40% acrylamide solution (dose not stated) within 42 min, and all animals eventually succumbed (Novikova, 1979). Although limited experimental data suggest that this compound may cause cancer and interfere with reproductive capacity, the effects of major concern are related to the neurotoxicity (WHO, 1985). The neurotoxic potential has been substantiated by long-term animal experiments using a variety of species, in particular rats and cats (McCollister et al., 1964; Novikova, 1979; Burek et al., 1980). These studies suggest that the no-effect level for neurotoxicity in the species studied would be somewhat below 1 mg kg^{-1} body weight/day.

Human data

Detailed data on skin penetration in humans are not available. Both the peripheral and the central nervous system may be involved, and a total of more than 50 cases of acrylamide-related neurotoxicity have been described in the medical literature (WHO, 1985; Auld and Bedwell, 1967; Garland and Patterson, 1967; Cavigneaux and Cabasson, 1971; Igisu et al., 1975; Mapp et al., 1977). Most cases of human acrylamide intoxication have occurred under circumstances where skin absorption was likely to occur, and intake through other routes was apparently minimal in several cases (Auld and Bedwell, 1967; Garland and Patterson, 1967; Cavigneaux and Cabasson, 1971; Takahashi et al., 1971; Igisu et al., 1975). Unfortunately, dose estimations have not been possible, and this fact is partly due to the lack of a method for biological monitoring. Thus, the contribution by skin absorption cannot be assessed in detail. Acrylamide may act as a skin irritant, and extensive skin contact has resulted in blistering and desquamation (Mapp et al., 1977). The onset of poisoning is usually insidious, and the intoxications reported followed an exposure period as short as 10 days, although a few months was more frequently the case. Dermatitis and central nervous system effects may be the first to occur, and the acutely poisoned patient typically exhibits a syndrome resembling a toxic psychosis with confusion, memory and concentration difficulties, and hallucinations. The peripheral effects are usually the result of chronic exposures to low-level exposures and appear with a latency of several weeks: muscular weakness, paraesthesia, loss of tendon reflexes and electrophysiological signs of polyneuropathy. These changes are slow to disappear, but total recovery seems to be possible even in severe cases. Taking differences in metabolic rate into consideration, experimental studies suggest a no-effect level for humans below 20 mg of acrylamide per day. However, this level does not take more subtle neurotoxic effects and individual susceptibility into consideration, and a safety factor should therefore be applied. Thus, a daily absorption in the order of magnitude of 1 mg could constitute a hazard (WHO, 1985). This compound is considered a possible carcinogen (class 2B) (IARC, 1987).

Conclusions

Acrylamide may penetrate the skin in significant quantities, and dermal absorption of this compound has caused or contributed to numerous intoxications. Acrylamide should therefore be regarded as a skin absorption hazard.

Acrylonitrile

Skin exposure potential

Acrylonitrile (2-propenenitrile) is usually produced by ammoxidation of propylene. The total production of acrylonitrile by 11 producers in the EC countries in 1983 was about 925 000 tonnes which almost covered the consumption within the community (ECDIN, 1987). Although used in the past as a pesticide, acrylonitrile is now mainly used for the production of polymers and co-polymers (acrylic and modacrylic fibres). In the USA where the amount of acrylonitrile production and use appears to be somewhat similar to that in the EC, about 12 000 workers come into major contact with acrylonitrile, and about 10-fold that number are also exposed to some extent (WHO, 1983). With decreasing exposure limits for acrylonitrile, both the number of workers with major exposure and the potential of significant skin exposure may have decreased. Contamination of the skin of workers, clothing, tools, equipment, etc., has been demonstrated and proved difficult to remove (WHO, 1983).

Physicochemical properties

At room temperature, acrylonitrile is an unstable liquid with a vapour pressure of 80 mm Hg. The solubility in water is about 7%, the octanol/water partition coefficient (log P_{ow}) is -0.92, and acrylonitrile is miscible with most organic solvents (WHO, 1983). The calculated solubility in psoriasis scales would suggest that skin uptake could occur, though not to a very large extent (Hansen and Andersen, 1988). Evaporation would limit absorption, and chemical reactions in the skin would cause local damage, thereby slowing down the percutaneous absorption.

Experimental data

LD_{50} values vary considerably with species and administration route. Dermal LD_{50} values using different animal models are about 200 mg kg^{-1} (Roudabush *et al.*, 1965), i.e. a few times higher than those seen with oral or intraperitoneal exposure. In a study of rabbits, the LD_{50} after skin application was 43 mg kg^{-1} (IARC, 1979). Percutaneous absorption of acrylonitrile vapour was demonstrated by Rogaczewska (1975) who exposed 315–350 cm^2 of the skin of rabbits (who breathed pure air) to a vapour concentration of 44–62 g m^{-3} and found that, although the penetration rate through skin was only about 1% of the rate of respiratory uptake, such skin exposure was fatal within 4 h. These data document the potential significance of skin absorption when respiratory protection alone is used to limit the uptake of this compound. In addition, local irritation and necrosis of the exposed skin have been reported (WHO, 1983). A range of toxic effects have been demonstrated in experimental studies, including carcinogenicity (WHO, 1983; Willhite, 1982).

Human data

Several studies have reported severe dermal effects of skin exposures to acrylonitrile (WHO, 1983). Also, in some cases of systemic toxicity, percutaneous absorption may have played an important role, although the available data do not allow a detailed evaluation. Most convincing is the report by Lorz (1950) of a 10-year-old girl who died after treatment of the scalp (for lice infestation) with an acrylonitrile-containing product. One study of absorption through human skin suggested a rate of $0.6\,\mathrm{mg\,cm^{-3}\,h^{-1}}$ (Rogaczewska and Piotrowski, 1968). Biological monitoring data of acrylonitrile and its metabolites show some correlation with respiratory exposures (Houthouijs et al., 1982), but the actual contribution of skin absorption in such studies has not been evaluated. Acrylonitrile may cause allergic contact dermatitis (Fisher, 1986). Acute effects are mainly non-specific and originate from the gastrointestinal and respiratory tracts, the liver and the central nervous system (WHO, 1983; Willhite, 1982). Epidemiological evidence is limited, and this compound is considered a probable carcinogen (class 2A) (IARC, 1979).

Conclusion

Acrylonitrile may penetrate the skin in significant quantities and may result both in local effects on the cutaneous tissues and in systemic toxicity. Acrylonitrile is a probable carcinogen. Acrylonitrile should therefore be regarded as a skin exposure hazard.

Other compounds

Skin penetration for other substances in this group is poorly documented. Tetramethylsuccinonitrile is of concern because of reproductive toxicity in experimental animals. Other nitriles, such as acetone cyanohydrin, isobutyronitrile and, apparently, adiponitrile, have caused systemic toxicity as a result of industrial accidents involving extensive dermal exposure (Zeller et al., 1969).

1-Butylamine is a primary irritant, and it has a dermal LD_{50} in the guinea pig of $366\,\mathrm{mg\,kg^{-1}}$; the oral LD_{50} in the rat is of similar magnitude (Smyth and Carpenter, 1944). Di-isopropylamine (bis(2-propyl)amine) is an irritant as well. The same is true for N,N-diethyl-2-aminoethanol (2-diethylaminoethanol) that has a dermal LD_{50} in the guinea pig of $884\,\mathrm{mg\,kg^{-1}}$, i.e. less than the oral LD_{50} in the rat (Smyth and Carpenter, 1944).

The epoxy resin curing agent, diethylenetriamine ((2-aminoethyl)-1,2-ethanediamine), is also a primary irritant; the dermal LD_{50} in the rabbit is $1.09\,mg\,kg^{-1}$, i.e. somewhat below the oral LD_{50} in rats (Smyth et al., 1949).

Systemic toxicity is important with *N*-nitrosodimethylamine (dimethylnitrosamine) because of its hepatotoxicity and carcinogenic potential (Cooper and Kimbrough, 1980). However, the dermal absorption potential is poorly documented.

References

ACGIH (1986), *Documentation for the threshold limit values and biological exposure indices*. 5th edn. (Cincinnati: American Conference of Governmental Industrial Hygienists).

AULD, R. B. and BEDWELL, S. F. (1967), Peripheral neuropathy with sympathetic overactivity from industrial contact with acrylamide, *Canadian Medical Association Journal*, **96**, 652-654.

BORM, P. J. A., DE JONG, L. and VLIEGEN, A. (1987), Environmental and biological monitoring of workers occupationally exposed to dimethylacetamide, *Journal of Occupational Medicine*, **29**, 898-903.

BORTSEVICH, S. V. (1984), The problem of the hygienic importance of dimethylformamide absorption through the skin (in Russian), *Gigiena Truda I Professionalaye Zaboleraniia*, **11**, 55-7.

BUREK, J. D., ALBEE, R. R., BEYER, J. E., BELL, T. J., CARREON, R. M., MORDEN, D. C., WADE, C. E., HERMANN, E. A. and GORZINSKI, S. J. (1980), Subchronic toxicity of acrylamide administered to rats in the drinking water followed up to 144 days of recovery, *Journal of Environmental Pathology and Toxicology*, **4**, 157-182.

CAMARASA, J. G. (1987), Contact dermatitis from dimethylformamide, *Contact Dermatitis*, **16**, 234.

CAVIGNEAUX, A. and CABASSON, G. B., (1971), Intoxication par l'acrylamide, *Archives des maladies professionelles de médicine du travail et de sécurité sociale*, **33**, 115-116.

COOPER, S. W. and KIMBROUGH, R. D. (1980), Acute dimethylnitrosamine poisoning outbreak, a case report, *Journal of Forensic Science,* **25**, 874-82.

CORSI, G. C. (1971), Sulla patologia professionale da dimetilacetamide (con particolare riferimento alla funzionalità epatica), Medicine del Lavoro, **67**, 28-41.

CZAJKOWSKA, T. (1981), Comparative evaluation of absorption of toxic compounds through the skin of rabbits and rats, *Przeglad Lekarski*, **38**, 659-62.

DUCATMAN, A. M., CONWILL, D. E. and CRAWL, J. (1986), Germ cell tumors of the testicle among aircraft repairmen, *Journal of Urology*, **136**, 834-6.

FISEROVA-BERGEROVA, V. and PIERCE, J.T. (1989), Biological monitoring V. Dermal absorption, *Applied Industrial Hygiene*, **4**, F14-F21.

FISHER, A. A. (1986), *Contact dermatitis*, 3rd edn. (Philadelphia, Lea & Febiger).

GARLAND, T. O. and PATTERSON, M. W. H. (1967), Six cases of acrylamide poisoning, *British Medical Journal*, **4**, 134-138.

HANSEN, C. M. and ANDERSEN, B. H. (1988), The affinities of organic solvents in biological systems, *American Industrial Hygiene Association Journal*, **49**, 301-8.

HASHIMOTO, K. and ANDO, K. (1975), Studies on the percutaneous absorption of acrylamide. *Abstracts of the XVIII International Congress on Occupational Health*, Brighton, p. 453.

HORN, H. J. (1961), Toxicology of dimethylacetamide, *Toxicology and Applied Pharmacology*, **3**, 12-24.

HOUTHUIJS, D., REMIJN, B., WILLEMS, H., BOLEIJ, J. and BIERSTEKER, K. (1982), Biological monitoring of acrylonitrile exposure, *American Journal of Industrial Medicine*, **3**, 313-20.
IARC (1979), *Some monomers, plastics and synthetic elastomers.* (IARC Monographs on the evaluation of the carcinogenic risk of chemicals to humans, Vol. 19) (Lyon: International Agency for Research and Cancer).
IARC (1987), *Overall evaluations of carcinogenicity: An updating of IARC Monographs Volumes 1 to 42*, (IARC Monographs on the evaluation of carcinogenic risk of chemicals to man (Suppl. 7.)) (Lyon: International Agency for Research on Cancer).
IARC (1989), *Some organic solvents, resin monomers and related compounds, pigments and occupational exposures in paint manufacture and painting* (IARC Monographs on the evaluation of carcinogenic risk to humans, Vol. 47) (Lyon: International Agency for Research on Cancer).
IGISU, H., GOTO, I., KAWAMURA, Y., KATO, M., IZUMI, K. and KUROIWA, Y. (1975), Acrylamide encephaloneuropathy due to well water pollution, *Journal of Neurology, Neurosurgery and Psychiatry*, **38**, 581-4.
KENNEDY, G. L. JR. (1986), Biological effects of acetamide, formamide,and their monomethyl and dimethyl derivatives, *CRC Critical Review of Toxicology*, **17**, 129-82.
KENNEDY, G. L. JR. and SHERMAN, H. (1986), Acute and subchronic toxicity of dimethylformamide and dimethylacetamide following various routes of administration, *Drug Chemistry and Toxicology*, **9**, 147-70.
KIMBROUGH, R. D., MAHAFFEY, K. R., GRANDJEAN, P., SANDOE, S. -H. and RUTSTEIN, D. D. (1989), *Clinical effects of environmental chemicals* (New York: Hemisphere).
LAUWERYS, R. R., KIVITS, A., LHOIR, M., RIGOLET, P., HOUBEAU, D., BUCHET, J. P. and ROELS, H. A. (1980), Biological surveillance of workers exposed to dimethylformamide and the influence of skin protection on its percutaneous absorption, *International Archives of Occupational and Environmental Health*, **45**, 189-203.
LORZ, H. (1950), Über perkutane Vergiftung mit Akrylnitril (Ventox) *Deutsche medizinische Wochenschrift*, **75**, 1087-8.
MAPP, C., MAZZOTTA, M., BARTOLUCCI, G. B. and FABRI, L. (1977), Neuropathy due to acrylamide: first observations in Italy (In Italian), *Medicina del Lavoro*, **68**, 1-12.
MASSMAN, W. (1956), Toxicological investigations on dimethylformamide, *British Journal of Industrial Medicine*, **13**, 51-4.
MAXFIELD, M. E., BARNES, J. R., AZAR, A. and TROCHIMOWICZ, H. T. (1975), Urinary excretion of metabolite following experimental human exposures to DMF and DMAC, *Journal of Occupational Medicine*, **17**, 506-11.
MCCOLLISTER, D. D., OYEN, F. and ROWE, V. K. (1964), Toxicology of acrylamide, *Toxicology and Applied Pharmacology*, **6**, 172-81.
MUKHTAR, H., DIXIT, R. and SETH, P. K. (1981), Reduction in cutaneous and hepatic glutathione contents, glutathione-S-transferase and aryl hydrocarbon hydroxylase activities following topical application of acrylamide to mouse, *Toxicology Letters*, **9**, 152-156.
MUNRO, D. D. and STOUGHTON, R. B. (1965), Dimethylacetamide (DMAC) and dimethylformamide (DMFA) effect on percutaneous absorption, *Archives of Dermatology*, **92**, 585-6.
NOVIKOVA, E. E. (1979), Toxic effect of acrylamide after entering through the skin (in Russian), *Gigienai Sanitarü*, **10**, 73-74.
PAOLETTI, A., FABRI, G. and BETTOLO, P. M. (1982), Un caso insolito di "addome acuto", Intosicazione da dimetilformamide, *Minerva Medica*, **73**, 3407-10.
POTTER, H. P. (1973), Dimethylformamide-induced abdominal pain and liver injury, *Archives of Environmental Health*, **27**, 340-1.

ROGACZEWSKA, T. (1975), Absorption of acrylonitrile vapours through the skin in animals (in Polish), *Medycyna Pracy*, **25**, 459-65.
ROGACZEWSKA, T. and PIOTROWSKI, J. (1968), Experimental evaluation of the absorption routes of acrylonitrile in humans (in Polish), *Medycyna Pracy*, **19**, 349-54.
ROUDABUSH, R. L., TERHAAR, C. J., FASSETT, D. W. and DZINBA, S. P. (1965), Comparative acute effects of some chemicals on the skin of rabbits and guinea pigs, *Toxicology and Applied Pharmacology*, **7**, 559-65.
SCAILTEUR, V. and LAUWERYS, R. R. (1987), Dimethylformamide (DMF) hepatotoxicity, *Toxicology*, **43**, 231-8.
SMYTH, H. F. and CARPENTER, C. P. (1944), The place of the range-finding test in the industrial toxicology laboratory, *Journal of Industrial Hygiene and Toxicology*, **26**, 269-73.
SMYTH, H. F., CARPENTER, C. P. and WEIL, C. S. (1949), Range-finding toxicity data, list III, *Journal of Industrial Hygiene and Toxicology*, **31**, 60-2.
SMYTH, H. F., CARPENTER, C. P., WEIL, C. S., POZZANI, U. C. and STRIEGEL, J. A. (1962), Range-finding toxicity data, list VI. *Amercian Industrial Hygiene Association Journal*, **28**, 95-107.
STULA, E. F. and KRAUSS, W. C. (1977), Embryotoxicity in rats and rabbits from cutaneous application of amide-type solvents and substituted ureas, *Toxicology and Applied Pharmacology*, **41**, 35-55.
TAKAHASHI, M., OHARA, T. and HASHIMOTO, K. (1971), Electrophysiological study of nerve injuries in workers handling acrylamide, *International Archiv für Arbeitsmedizin*, **28**, 1-11.
WHO (1983), *Acrylonitrile*, (Environmental health criteria 28) (Geneva: World Health Organization).
WHO (1985), *Acrylamide*, (Environmental health criteria 49) (Geneva: World Health Organization).
WILLHITE, C. C. (1982), Toxicology updates. Acrylonitrile, *Journal of Applied Toxicology*, **2**, 54-6.
ZELLER, H., HOFMANN, H. T., THIESS, A. M. and HEY, W. (1969), Zur Toxicität der Nitrile. *Zentralblatt für Arbeitsmedizin und Arbeitsschutz*, **19**, 225-38.

Chapter 6
Isocyclic hydrocarbons, alcohols and related compounds

Introduction

The isocyclic compounds most frequently considered to be skin penetrants are listed in table 6.1. The most important of these are described in the sections below.

Benzene and its derivatives would be likely to penetrate the skin in significant quantities, depending on their toxic potential; the polynuclear aromatic hydrocarbons would, because of their strong lipophilicity and large molecular size, penetrate more slowly.

Calculated solubility in psoriasis scales would indicate considerable dermal uptake of several aromatic compounds, perhaps most rapidly with dimethyl phthalate (Hansen and Andersen, 1988).

In addition to the compounds listed in table 6.1 cyclohexanone has received a 'skin' denotation in one country. Among the aromatic hydrocarbons with a 'skin' denotation in one or two countries are ethylbenzene, styrene (vinylbenzene), methylstyrene (2-propenylbenzene), naphthalene, benzo[a]pyrene, and coal tar. The two latter entries are probably considered cutaneous hazards because of their carcinogenicity (IARC, 1987). Some of these compounds can also cause allergic contact dermatitis (Fisher, 1986). The contribution by dermal uptake to systemic toxicity for these compounds is unclear.

Table 6.1. Isocyclic hydrocarbons, alcohols and related compounds that are considered a skin hazard in several countries (y = yes, n = no, o = other regulation). Those that are described in separate sections are italicized.

CAS No.	Chemical	Number of countries	FRG	Sweden	USA
71-43-2	*Benzene*	12	y	y	y
108-88-3	*Toluene*	6	n	y	n
98-82-8	*Cumene (2-propylbenzene)*	8	y	y	y
1330-20-7	*Xylenes*	10	o	y	n
108-95-2	*Phenol*	15	y	y	y
1319-77-3	*Cresols*	14	y	y	y
583-60-2	*2-Methylcyclohexanone*	11	y	n	y

With a series of substituted phenols, a close correlation was obtained between the permeability coefficient and the octanol/water partition coefficient, log P_{ow} (Roberts et al., 1977). Thus, the more lipophilic phenolic compounds would pass the skin more rapidly. 2-sec-butylphenol, and 4-tert-butylphenol belong to this group; they have a 'skin' denotation in one or two countries.

Phenyl ether (diphenyl ether), and benzaldehyde are also considered a skin hazard in one or two of the countries surveyed. Cyclohexanone peroxide, dicyclohexyl peroxide, benzoyl peroxide (dibenzoylperoxide), and *alpha, alpha*-dimethylbenzyl hydroperoxide have a 'skin' denotation in single countries, probably because of the skin irritation or risk of sensitization (Fisher, 1986).

Pyrethrum is a mixture of compounds and has been listed in one country as a percutaneous absorption hazard; contact dermatitis due to this product is relatively frequent (Fisher, 1986).

Benzene

Skin exposure potential

Benzene is widely produced from crude oil with an annual production in EC countries of about 35 million tonnes (in 1980) (ECDIN, pers. comm. 1987); benzene is produced mainly from pyrolysis of gasoline, while coke-oven operations provide less than 10% of the total production (CEFIC, 1983). Most of the benzene is used in the production of substituted aromatic compounds, some for production of detergents, pesticides and solvent mixtures, and a large amount is present in petrol used by automobiles. The major skin exposure hazard occurs in the user trades, such as tyre manufacture, printing and painting, and other processes where benzene-containing thinners and degreasers are used. In the USA, 1.5 million individuals are exposed to benzene, 14% having direct occupational exposure (NIOSH, 1976a), possibly involving skin contact. The exposure potentials have decreased during recent years due to the discouragement of benzene usage as a thinner. Recently, a questionnaire study has indicated that a minimum of 13 000 workers may be exposed to benzene in EC countries (CEFIC, 1983).

Physicochemical properties

At room temperature, benzene is a liquid with a vapour pressure of 75 mm Hg. The solubility in water is only 0.06%, and benzene is miscible with ethanol, diethyl ether, toluene and other organic solvents. The octanol/water partition coefficient (log P_{ow}) has been reported as 2.03–2.15. Blank and McAuliffe (1985) measured partition coefficient for benzene using human skin (stratum corneum) vs different vehicles; the stratum corneum/water partition coefficient was 30, i.e. about 100-fold above those seen with stratum corneum and petrol, hexane and other solvents. Furthermore, 1.5 mg of powdered stratum corneum from human calluses absorbed about one-sixth of the benzene present in a 1.5 ml aqueous solution (Wester et al., 1987). A high solubility was also found in psoriasis scales (Hansen and Andersen, 1988). On the basis of physicochemical properties, a penetration rate of $0.70\,mg\,cm^2\,h^{-1}$ was calculated by Fiserova-Bergerova and Pierce (1989).

Experimental data

The penetration rate of benzene through excised rat skin was $0.19\,mg\,cm^{-2}\,h^{-1}$ (Tsuruta, 1982). Peak recovery of benzene in exhaled air can be demonstrated within the first 15 min after skin exposure (Susten et al., 1985), thus confirming the rapid penetration. However, due to the volatility of benzene, only about 1% of the amount applied, whether undiluted or present in a solvent in a low concentration, was absorbed by hairless mice (Susten et al., 1985). In rhesus monkeys, similar or slightly lower absorptions were obtained when the benzene was applied to the palmar surface, to the forearm skin after removal of the stratum corneum (by stripping), or after multiple washes with benzene; much lower absorption was seen with undamaged forearm skin and when benzene occurred in a 0.35% concentration in a rubber solvent (Maibach and Anjo, 1981). *In vitro* studies using human skin documented a 100-fold higher permeability constant for benzene in water as compared to benzene in petrol or other organic solvents; also, penetration of benzene vapour through the skin was documented (Blank and McAuliffe, 1985). Local damage to the skin is well known and has been studied in detail by Wohlrab and Wozniak (1984). This tissue damage may result in increased percutaneous uptake of other compounds (Blank and McAuliffe, 1985).

Human data

Early studies by Hanke et al. (1961) suggested a maximal rate of skin absorption in humans (with occlusion) at about $0.4\,mg\,cm^{-2}\,h^{-1}$; the exposed skin on the forearm was occluded for up to 2 h, and the methods used would suggest that the absorption rate was somewhat overestimated. In the absence of other and more exact information, Blank and McAuliffe (1985) extrapolated from their *in vitro* studies of human skin and concluded that, with $100\,cm^2$ of skin in contact with a 5% benzene solution, skin absorption would be comparable to the respiratory uptake from a benzene concentration of 10 p.p.m. in the air. Other calculations suggested that exposure of a skin area of $360\,cm^2$ to benzene would cause an absorption more than 30-fold above inhalation of benzene at the TLV level for the same time period (Fiserova-Bergerova and Pierce, 1989). However, from information on skin exposures to rubber solvents in tyre-building operations, Susten et al. (1985) calculated a lower skin absorption which would be less than the respiratory absorption from 1 p.p.m. in the air. Acute exposure to benzene may result in central nervous system depression and other systemic toxicity, while chronic exposures are mainly related to aplastic anaemia and leukaemia (NIOSH, 1976). Benzene is regarded as a human carcinogen (class 1) (IARC, 1987). In the recent study by Rinsky et al. (1987), the cumulative benzene exposure in rubber work was associated with the leukaemia risk, and long-term exposure to 1 p.p.m. of benzene caused a statistically significant risk. Although skin absorption was not addressed in this study, the epidemiological evidence suggests that the total benzene absorption should be minimized.

Conclusion

Benzene may penetrate the skin in significant quantities, but absorption is limited by evaporation and may also be decreased if benzene occurs in a mixture of solvents. Repeated skin exposure may damage the skin. The most important systemic effect is leukaemia. Thus, benzene should be regarded as a skin exposure hazard.

Toluene

Skin exposure potential

Toluene (methylbenzene) is primarily produced by catalytic reforming of naphtha, and technical toluene usually contains benzene, xylenes and several other impurities. Total production within the EC was 1 million tonnes in 1980, an amount similar to the total consumption. Toluene, as a component of reformate, is primarily used in petrol, and, as a component of white spirit, in solvents. Isolated toluene is also used as a solvent and a petrol additive, and in the production of benzene, toluene di-isocyanate and other aromatic compounds. Occupational exposures occur whenever toluene or toluene-containing solvents and thinners are used for rubber, resins, asphalt, paints, varnishes, photogravure, and for cleaning and defatting. Skin exposures occur in particular when the solvent is used, e.g. for cleaning paint off the hands. An estimated 1 278 000 workers in the USA are potentially exposed to toluene (IARC, 1989).

Physicochemical properties

This aromatic compound is a liquid at room temperature, and has a vapour pressure of 28 mm Hg at 25°C; the boiling point is 110°C. Toluene is soluble in acetone and carbon disulfide, it is miscible with most other organic solvents, such as benzene, ethyl alcohol and ethyl ether, and 1 l of water will dissolve about 0.5 g of toluene; the octanol–water partition coefficient (log P_{ow}) has been reported as 2.11-2.80 (NAS, 1981). The molecular weight is 92. The limited solubility of toluene in psoriasis scales predicts a poor uptake in the skin (Hansen and Andersen, 1988). Assuming no toxic effect on the stratum corneum, a penetration rate of 0.69 mg cm^{-2} h^{-1} was calculated by Fiserova-Bergerova and Pierce (1989) on the basis of physicochemical properties.

Experimental data

LD$_{50}$ values in the rabbit are 14 g kg^{-1} for cutaneous exposure and 7.5 g kg^{-1} for oral administration (Smyth et al., 1969). An in vitro study using excised abdominal skin of the rat reported that steady-state conditions were obtained after a lag time of about 90 min; the penetration rate was 47 µg cm^{-2} h^{-1}, i.e., one-fourth of that seen with benzene and almost 10-fold above that seen with o-xylene (Tsuruta, 1982). The in vivo percutaneous absorption rate in mice was 4.59 µg cm^{-2} h^{-1} (Tsuruta et al., 1987). The permeability constant k_p for toluene applied to human skin in vitro was 8.1×10^4 cm h^{-1}, but it increased to 210×10^4 cm h^{-1} when applied in an ethanol:water (1:1) solution; in the latter case, the lag time was only 3–4 min (Dugard and Scott, 1984). In a study of guinea pigs, dermal application of 1 ml toluene on 3.1 cm^2 of skin under occlusion caused a rapid increase in blood toluene concentration during the first hour, followed by a decrease despite continued dermal exposure (Jakobson et al., 1982). Blood concentrations of up to 15 µmol l^{-1} (1.4 mg l^{-1}) were obtained by Boman (1989) during a 6-h exposure to toluene on a skin area of 3.14 cm^2 in guinea pigs. Prior stripping of the stratum corneum, abrasion by sandpaper or scratching with a needle resulted in lower blood concentrations during subsequent dermal exposure, presumably because of removal of lipophilic structures of the stratum corneum (Boman, 1989). In addition to neurotoxicity, animal experiments have indicated that toluene may cause immunotoxicity and hepatotoxicity (Jelnes, 1989).

Human data

In human volunteers, the immersion of one hand for 5 min in toluene resulted in considerably increased toluene levels in blood taken from the same arm, as compared to the other arm, and this difference lasted for 3 h (Aitio et al., 1984). Up to 0.2–0.5 mg l^{-1} was found in the blood of the contralateral arm after a 5-min immersion of one hand in toluene, and these levels were almost as high as the blood levels achieved from inhalation of 100 p.p.m. toluene at rest (Aitio et al., 1984). If one hand of a volunteer was immersed in toluene for 1 min, hardly any toluene could be detected in the blood (Sato and Nakajima, 1978). However, after a 30-min immersion, the average blood level in 5 volunteers reached almost 0.2 mg l^{-1}, about one-quarter of that achieved after inhalation of 100 p.p.m. vapour for 4 h (Sato and Nakajima, 1978). Similar results were obtained in a third study, but one subject showed a maximum of 3.01 mg l^{-1}, 2 h after a simulated rinsing of one hand (Christensen, 1982). Such high levels may be due to increased absorption through abrasions and cuts of the skin. The maximal absorption rate of toluene into the skin in nine human volunteers was estimated by the direct method as 14–23 mg cm^{-2} h^{-1} (Dutkiewicz and Tyras, 1968a), but the true absorption rate would be much less, as steady state was not achieved. The absorption of toluene from an aqueous solution is proportional to the toluene concentration and was found to reach 0.6 mg cm^{-2} h^{-1} for the saturated solution (Dutkiewicz and Tyras, 1968b). Skin contract for 30 min results only in a mild, burning sensation and a slight erythema (Sato and Nakajima, 1978). Thus, considerable variation in skin penetration rates occurs, but skin absorption of liquid toluene seems to be of major significance when pulmonary exposures are kept below 100 p.p.m. Theoretical calculations support this impression (Fiserova-Bergerova and Pierce, 1989). The percutaneous absorption of toluene vapours has been assessed for a 600 p.p.m. exposure during 3.5 h; the total skin absorption of the vapours contributed less than 1% of the concomitant pulmonary absorption (Riihimäki and Pfäffli, 1978). Thus, only skin contact with the liquid solvent contributes a significant absorption, and penetration of toluene through the skin is a relatively slow and prolonged process. Acute symptoms of toluene toxicity are mainly from the central nervous system: headache, nausea, dizziness and drowsiness; during longer-term exposure, the same symptoms may become more chronic, though decreasing in severity or disappearing during week-ends (National Academy of Sciences, NAS, 1981). Chronic toluene exposure is suspected to be a contributing cause of chronic neurotoxicity which clinically appears as toxic encephalopathy (Jelnes, 1989; NAS, 1981).

Conclusion

Toluene may penetrate the skin in significant amounts, although the penetration process appears to be prolonged. Toluene should therefore be regarded as a skin absorption hazard.

Xylenes

Skin exposure potential

Xylenes (dimethylbenzenes, three isomers) are mainly produced by catalytic reforming of naphtha streams and subsequent separation. The xylene isomers differ only slightly in their properties and occur as a mixture in commercial xylene, the *meta* isomer constituting the main part. Ethyl benzene and other aromatic impurities are usually present as well. Production of xylene within the EC was 1.2 million tonnes in 1980. It is used as a thinner for paints and varnishes, in pharmaceutical preparations, as an octane-booster in petrol, and in the synthesis of various organics. Skin exposure occurs in a wide range of operations. An estimated 1 106 000 workers are potentially exposed to xylenes in the USA (IARC, 1989).

Physicochemical properties

This colourless liquid has an aromatic odour and a vapour pressure of about 10 mm Hg at 30°C. The boiling points for the three isomers are about 140°C. Xylene is almost insoluble in water, only 0.13 g being dissolved in 1 l of water at

20°C, while it is very soluble in ethyl alcohol, diethyl ether and other organic solvents. The octanol/water partition coefficients (log P_{ow}) are 3.12, 3.20 and 3.15 for the *ortho*, *meta* and *para* isomers, respectively. The solubility of xylene in psoriasis scales is limited and suggests that uptake in the skin will not be extensive (Hansen and Andersen, 1988). On the basis of physicochemical properties and assuming no toxic effect on the stratum corneum, a penetration rate of 0.50 mg cm^{-2} h^{-1} has been calculated (Fiserova-Bergerova and Pierce, 1989).

Experimental data

In an experimental study using excised abdominal skin from the rat, a steady-state penetration rate for *o*-xylene was measured at 6 µg cm^{-2} h^{-1} after a lag time of 2 h, i.e. about one-tenth of the rate of toluene and one-hundredth of that of benzene (Tsuruta, 1982). Xylene is neurotoxic (ACGIH, 1986; NAS, 1981; Izmerov, 1984).

Human data

In 10 human volunteers, the rate of absorption into the skin was found by the direct method to vary between 4.5 and 9.6 mg cm^{-2} h^{-1} (Dutkiewicz and Tyras, 1968). Indirect assessment of percutaneous absorption by determination of excretion levels in 13 human volunteers varied between 42 and 260 µg cm^{-2} h^{-1} for *m*-xylene (Lauwerys *et al.*, 1978). This somewhat lower rate compares favourably with another study using the indirect method which showed an average absorption rate of 120 µg cm^{-2} h^{-1} (Engström *et al.*, 1977). The higher absorption rate suggested by the direct method (Dutkiewicz and Tyras, 1968b) may be caused by differences in the calculation under non-steady-state conditions. Continued systemic absorption has been documented for 5 h after skin exposure (Engström *et al.*, 1977). Higher absorption rates have occurred in an individual with a history of atopic dermatitis (Engström *et al.*, 1977), and even higher rates would be expected in cases of eczema or other skin diseases or lesions. Immersion of both hands in xylene for 15 min resulted in a systemic absorption of about 35 mg, an amount similar to the pulmonary absorption during the same time period when inhaling a xylene vapour concentration of 100 p.p.m. (Engström *et al.*, 1977). Thus, although skin exposures will generally be of rather short duration and only involve small skin areas, current skin exposures could add significantly to the total amount absorbed. Calculated contributions from respiratory and dermal exposures support this conclusion (Fiserova-Bergerova and Pierce, 1989). In addition, xylene-induced irritation and vasodilation may enhance the absorption rate. Acute exposures to xylene appear to cause dizziness, disorientation, dyspnoea and nausea, in severe cases progressing to coma; chronic exposures are associated with similar, non-specific symptoms and irritability, insomnia and memory loss (ACGIH, 1986; NAS, 1981; Izmerov, 1984). Xylene exposure may possibly contribute to cases of chronic encephalopathy induced by solvents.

Conclusion

Xylene may penetrate the skin in significant quantities, although rather slowly, and extensive skin exposure could result in systemic toxicity. Xylene should therefore be regarded as a skin absorption hazard.

Phenol

Skin exposure potential

Phenol (hydroxybenzene) is primarily produced from cumene (isopropylbenzene). Production of phenolic resins constitutes its major use, but phenol is also used for production of bisphenol A and several other organics; phenol has limited medical applications today. The phenol production within the EC was more than one million tonnes in 1983, and most was consumed within the community. In the USA, an estimated 193 000 workers are exposed to phenol (IARC, 1989). Intermittent exposure and skin contact may occur in a wide variety of occupations with a substantial number of employees.

Physicochemical properties

Phenol is a white, crystalline solid at room temperature. Phenol liquifies on absorption of water from air; the melting point is 41°C. The vapour pressure is 0.3 mm Hg at room temperature. The water solubility is about 9%, and phenol is also soluble in organic solvents. The log octanol/water partition coefficient (log P_{ow}) is about 1.48–1.51 (Roberts *et al.*, 1977). The partition coefficient for phenol between stratum corneum and water has been found to be about 5 (Andersen *et al.*, 1976). The calculated solubility in psoriasis scales suggests that skin penetration could be rapid (Hansen and Andersen, 1988). Assuming no interference with the stratum corneum barrier, a penetration rate of 4.62 mg cm^{-2}h^{-1} was calculated on the basis of physicochemical properties (Fiserova-Bergerova and Pierce, 1989).

Experimental data

The LD$_{50}$ values obtained in rats (669 mg kg^{-1}) (Conning and Hayes, 1970) and rabbits (850 mg kg^{-1}) (Vernot *et al.*, 1977) are of the same order of magnitude as those seen with other administration routes. An *in vitro* study using human abdominal stratum corneum showed significant variation of the phenol flux from a 1% aqueous solution, the average being 1.5 µg cm^{-2}h^{-1}; also the lag time for skin penetration varied considerably from the average of about 30 min (Southwell *et al.*, 1984). With a similar experimental procedure, the permeability coefficient for low concentrations of phenol in water was found to be 1.37×10^{-4} cm min^{-1}; due to tissue damage at concentrations above 1.5%, the permeability increased markedly (Roberts *et al.*, 1977). The skin damage showed a maximal response 1 h after application on murine skin, and it was unaffected by the type of vehicle used (Patrick et al., 1985). This action is most likely to be the reason why phenol acts as a tumour promoter, but not as an initiator, in studies of experimental skin cancer (Boutwell and Bosch, 1959). Systemic absorption of phenol results in general tissue damage due to the denaturation of proteins; poisoning affects all organ systems (NIOSH, 1976 a and b; Bruce et al., 1987).

Human data

Numerous cases of phenol poisoning have occurred as a result of percutaneous absorption, and fatalities have resulted within 10–15 min after skin exposure (NIOSH, 1976). Many cases were due to accidents, but adverse effects have also been seen after skin contact with phenol-containing resins or aqueous solutions (Hinkel and Kintzel, 1968). The clinical picture may develop rapidly and include shock, collapse, coma, convulsions and cyanosis in addition to the local skin damage (NIOSH, 1976; Bruce et al., 1987). The amount absorbed by humans from (non-damaging) aqueous solutions depends on the concentration and the duration of skin exposure; a crude estimate of the flux for phenol from a 30-min application of 1% solution in water is $0.3 \, \text{mg cm}^{-2} \text{h}^{-1}$ (Baranowska-Dutkiewicz, 1981). This estimate is much higher than those reported from *in vitro* studies. Evidence has also been published that phenol vapour may cause significant uptake through the skin (Piotrowski, 1971). Skin contact with phenol may result in severe skin damage and even gangrene (Abraham, 1972). Due to the low vapour pressure at low temperatures, respiratory exposures at the workplace would tend to be limited, thus adding to the relative significance of percutaneous absorption.

Conclusions

Phenol may penetrate the skin in significant quantities. Skin exposure to phenol has caused numerous cases of poisoning, several of them being rapidly fatal. Phenol should therefore be regarded a skin absorption hazard.

Cresols

Skin exposure potential

Cresols (methylphenols) occur as three different isomers in commercial cresol. The main uses are as antioxidants, wire enamel solvents, intermediates in the production of phenolic resins, phosphate esters and agricultural chemicals, and as solvents for other purposes. An estimated 11 000 individuals in the USA are occupationally exposed to cresol, although many more may be exposed intermittently from degreasing agents containing cresol (NIOSH, 1978).

Physicochemical properties

Only m-cresol is a liquid at room temperature (melting point, 12°C), while both o-cresol and p-cresol are solids (melting points, 31°C and 35°C, respectively). Accordingly, the vapour pressures at ambient temperatures are relatively low. Cresol is soluble in caustic alkalies and miscible with alcohol, glycerol, benzene, ether and other organic solvents (NIOSH, 1978). The solubility in water is about 2%. The octanol/water partition coefficients (log P_{ow}) are similar for the three isomers, i.e. about two (Roberts et al., 1977). The partition coefficient for cresols between stratum corneum and water has been found to be about 10 (Andersen et al., 1976). The solubility of m-cresol in psoriasis scales suggests that uptake in the skin could be considerable (Hansen and Andersen, 1988).

Experimental data

Permeability coefficients for the three isomers in low concentrations were about $2.5-3.0 \times 10^{-4}$ cm min^{-1} in a study using epidermis from human abdominal skin (Roberts et al., 1977). At higher concentrations of cresol (about 1% for o-cresol and m-cresol; and almost 9% for p-cresol), the skin penetration increases markedly, thus indicating damage to the skin barrier (Roberts et al., 1977). Dermal LD$_{50}$ values were 1.4, 2.1 and 0.3 mg kg^{-1} for three isomers in rabbits, and 1.8 mg kg^{-1} for a mixture of these isomers in the same species (NIOSH, 1978). Skin irritation and erythema were seen in the exposed animals. These LD$_{50}$ values were somewhat above those seen by other administration routes. A tumour promotion study resulted in a high rate of skin papillomas related to repeated cresol application after initial skin exposure to dimethylbenzanthracene (Boutwell and Bosch, 1959).

Human data

A man who worked with his hands immersed most of the time for 5-6 h in a solution of 6% cresylic acid (with a small, though unknown, concentration of cresol) suffered severe skin damage, impaired speech and facial paralysis (Klinger and Norton, 1945). More convincing is the report of a male infant who had about 20 ml of a 90% cresol solution in water accidentally poured over his head; within 5 min, the baby was unconscious and cyanotic, and died 4 h later (Green, 1975). Skin absorption is otherwise poorly documented, but cresol-related chemical burns and dermatitis have been frequently reported; allergic contact dermatitis also occurs (Fisher, 1986). The systemic effects include haemolysis and bleeding, in particular in the liver and kidneys with oliguria or anuria (NIOSH, 1978).

Conclusions

Cresols may penetrate the skin in significant quantities and may cause corrosive damage to the skin. Although the documentation is limited, the properties of cresol and the systemic toxicity would suggest that cresol should be regarded as a skin exposure hazard.

Other compounds

Styrene and ethylbenzene were estimated to be potential skin absorption hazards, with calculated penetration rates of 0.52 and 0.53 mg cm^{-2}h^{-1}, respectively; these rates were considered significant in comparison with respiratory uptake (Fiserova-Bergerova and Pierce, 1989). In an experimental study using excised abdominal skin from the rat, a steady-state penetration rate for styrene and ethylbenzene was found to be about 30 and 6 µg cm^{-2}h^{-1} after a lag time of about 2 h (Tsuruta, 1982). The absorption rate for styrene in human volunteers was approximately 0.06 mg cm^{-2}h^{-1}, as determined from excreted and exhaled metabolites; percutaneous absorption resulted in much lower styrene uptake than did inhalation of this compound (Berode et al., 1985). With ethylbenzene, the percutaneous absorption rate was estimated at as much as 28 mg cm^{-2}h^{-1} by the direct method, and about 2 mg cm^{-2}h^{-1} when calculated from urinary excretion of mandelic acid in human volunteers; the maximum urinary concentrations were seen within 2 h after termination of the dermal exposure (Dutkiewicz and Tyras, 1967).

Among several isocyclic compounds, the rate of skin penetration in rats decreased in the following order: isopropylbenzene, p-menthene = myrcene, cyclohexane, ethylbenzene; the penetration time was between 20 and 31 min (Valette and Cavier, 1954).

2-Methylcyclohexanone has a dermal LD$_{50}$ of 1.77 ml kg^{-1} (1635 mg kg^{-1}) in rabbits, i.e. similar to the value obtained by oral exposure (Smyth et al., 1969). Thus, although of limited toxicity, this compound appears to penetrate the skin quite readily. Cumene is a primary irritant, and no documentation is available to suggest that systemic toxicity can occur from skin contact with this compound.

Topical application of coal tar for treatment of skin diseases has resulted in percutaneous absorption of acridine (Cernikova et al., 1983) and pyrene (Jongeneelen et al., 1985), as indicated by analysis of urine samples. Other patients showed increased urine mutagenicity (Wheeler et al., 1981). Workers with cutaneous exposure to creosote, a coal tar distillation product used for impregnation of wood, also showed increased excretion of 1-hydroxypyrene, a metabolite of pyrene (Jongeneelen et al., 1985).

Following percutaneous application of naphthalene at a dose of 0.5 mg kg^{-1} body weight, a maximum urinary excretion of 1-naphthol at 3–6 µmol l^{-1} was detected with a delay of about 4 h; the urinary levels were about one-tenth of the maximum levels associated with exposures to airborne naphthalene concentrations of about 2–4 mg m^{-3}. The percutaneous absorption was estimated to be approximately 3–10% of the dose (Luotamo et al., 1989). Naphthalene has a solubility in psoriasis scales that would suggest a relatively rapid dermal uptake (Hansen and Andersen, 1988).

^{14}C-Labelled benzo[a]pyrene and 7,12-dimethylben[a]anthracene were applied to the skin of mice and appeared to be rapidly absorbed; within 7 days, about 90% of the compounds were recovered in the excreta (Sanders et al., 1986). After dermal absorption, aryl hydrocarbon hydroxylase in the skin may convert non-carcinogenic benzo[a]pyrene into a carcinogen that could cause effects within the skin or elsewhere in the body (Wester and Maibach, 1984; IARC, 1987).

Hydroquinone (1,4-dihydroxybenzene) penetrates human skin; after application on the forehead, peak absorption occurred within 12 h, and about half of the amount was absorbed, depending on the hydroquinone preparation used (Bucks et al., 1988).

References

ABRAHAM, A. J. (1972), A case of carbolic acid gangrene of the thumb, *British Journal of Plastic Surgery,* **25,** 282–4.
ACGIH (1986) *Documentation for the threshold limit values and biological exposure indices,* 5th edn (Cincinnati: American Conference of Governmental Industrial Hygienists).
AITIO, A., PEKARI, K. and JÄRVISALO, J. (1984) Skin absorption as a source of error in biological monitoring, *Scandinavian Journal of Work and Environmental Health,* **10,** 317–320.
ANDERSEN, R.A., TRIGGS, E.J., and ROBERTS, M.S. (1976), The percutaneous absorption of phenolic compounds, 3. Evaluation of permeability through human stratum corneum using a desorption technique, *Australian Journal of Pharmaceutical Science,* **5,** 107–10.
BARANOWSKA-DUTKIEWICZ, B. (1981), Skin absorption of phenol from aqueous solutions in men, *International Archives of Occupational and Environmental Health,* **49,** 99–104.
BERODE, M., DROZ, P.O. and GUILLEMIN, M. (1985), Human exposure to styrene VI. Percutaneous absorption in human volunteers, *International Archives of Occupational and Environmental Health,* **55,** 331–6.
BLANK, I.H. and MCAULIFFE, D.J. (1985), Penetration of benzene through human skin, *Journal of Investigative Dermatology,* **85,** 522–6.
BOMAN, A. (1989), *Factors influencing the percutaneous absorption of organic solvents* (Arbete och Hälsa Vol 1989: 11) (Stockholm: National Institute of Occupational Health).
BOUTWELL, R.K. and BOSCH, D.K. (1959), The tumor-promoting action of phenol and related compounds for mouse skin, *Cancer Research,* **19,** 413–24.
BRUCE, R.M., SANTODONATO, J. and NEAL, M.W. (1987) Summary review of the health effects associated with phenol, *Toxicology and Industrial Medicine,* **3,** 535–68.
BUCKS, D.A.W., MCMASTER, J.R., GUY, R.H. and MAIBACH, H.I. (1988), Percutaneous absorption of hydroquinone in humans: Effect of 1-dodecylazacyclo-heptan-2-one (Azone) and the 2-ethylhexyl ester of 4-(dimethylamino)benzoic acid (Escalol 507), *Journal of Toxicology and Environmental Health,* **24,** 279–89.
CEFIC (1983), *Criterium document on benzene* (Brussels: European Council of Chemical Manufacturers' Federation).
CERNIKOVA, M., DUBSKI, H. and HORACEK, J. (1983), Detection of acridine in human urine after topical coal-tar treatment, *Journal of Chromatography,* **273,** 202–6.
CHRISTENSEN, J.M. (1982), Danish National Institute of Occupational Health, Unpublished results.

CONNING, D.M. and HAYES, M.J. (1970), The dermal toxicity of phenol, an investigation of the most effective first-aid measures, *British Journal of Medicine,* **27**, 155-9.
DUGARD, P.H. and SCOTT, R.C. (1984), Absorption through the skin. In *The Chemotherapy of Psoriasis,* (International Encyclopedia of Pharmacology and Therapeutics, Section 110), ed. H.P. Baden (Oxford: Pergamon).
DUTKIEWICZ, T. and TYRAS, H. (1967), A study of the skin absorption of ethylbenzene in man, *British Journal of Industrial Medicine,* **24**, 330-2.
DUTKIEWICZ, T. and TYRAS, H. (1968a), The quantitative estimation of toluene skin absorption in man, *International Archives Gewerbepathologie Gewerbehygiene,* **24**, 253-257.
DUTKIEWICZ, T. and TYRAS, H. (1968b), Skin absorption of toluene, styrene, and xylene by man, *British Journal of Industrial Health,* **25**, 243.
ENGSTRÖM, K., HUSMAN, K. and RIIHIMÄKI, V. (1977), Percutaneous absorption of m-xylene in man, *International Archives of Occupational Environmental Health,* **39**, 181-189.
FISEROVA-BERGEROVA, V. and PIERCE, J.T. (1989), Biological monitoring. V. Dermal absorption, *Applied Industrial Hygiene,* **4**, F14-F21.
FISHER, A.A. (1986), *Contact dermatitis,* 3rd edn. (Philadelphia, Lea & Febiger).
GREEN, M.A. (1975), A household remedy misused — Fatal cresol poisoning following cutaneous absorption, *Medical Science and Law,* **15**, 65-6.
HANKE, J., DUTKIEWICZ, T., and PIOTROWSKI, J. (1961), The absorption of benzene through the skin in man (in Polish), *Medyayna Pracy,* **12**, 413-26.
HANSEN, C.M. and ANDERSEN, B.H. (1988), The affinities of organic solvents in biological systems, *American Industrial Hygiene Association Journal,* **49**, 301-8.
HINKEL, G.K. and KINTZEL, H.W. (1968), Phenolvergiftigung bei Neugeborenen durch kutane Resorption, *Deutsche Gesundheitswesen,* **23**, 2420-2.
IARC (1987), *Overall evaluations of carcinogenicity: An updating of IARC Monographs Volumes 1 to 42,* (IARC Monographs on the evaluation of carcinogenic risk of chemicals to man (Suppl. 7)) (Lyon: International Agency for Research on Cancer).
IARC (1989), *Some organic solvents, resin monomers and related compounds, pigments and occupational exposures in paint manufacture and painting,* (IARC Monographs on the evaluation of carcinogenic risks to humans, Vol. 47). (Lyon: International Agency for Research on Cancer).
IZMEROV, N.F., ed. (1984), *Xylene* (Scientific reviews of Soviet literature on toxicity and hazards of chemicals 52) (Moscow: Centre of International Projects).
JAKOBSON, I., WAHLBERG, J.E., HOLMBERG, B. and JOHANSSON, G. (1982), Uptake via the blood and elimination of 10 organic solvents following epicutaneous exposure of anesthetized guinea pigs, *Toxicology and Applied Pharmacology,* **63**, 181-7.
JELNES, J.E., (1989), Toluene (in Danish). (Nordic expert group for documentation on occupational exposure limits, No. 82), *Arbete och Hälsa,* **3**, 53 pp.
JONGENEELEN, F.J., ANZION, R.B.M., LEIJDEKKERS, C.-M., BOS, R.P. and HENDERSON, P.T. (1985), 1-Hydroxypyrene in human urine after exposure to coal tar and a coal tar derived product, *International Archives of Occupational and Environmental Health,* **57**, 47-55.
KLINGER, M.E. and NORTON, J.F. (1945), Toxicity of cresylic acid-containing solvent, *US Naval Medicine Bulletin,* 438-9.
LAUWERYS, R.R., DATH, T., LACHAPELLE, J.-M., BUCHET, J.-P. and ROELS, H. (1978), The influence of two barrier creams on the percutaneous absorption of m-xylene in man, *Journal of Occupational Medicine,* **20**, 17-20.
LUOTAMO, M., HEIKKILÄ, P., RIIHIMÄKI, V. and ROMO, M. (1989), Urinary 1-naphthol as an indicator of exposure to naphthalene and creosote (poster). V. International Congress on Toxicology, Brighton, 17-21 July 1989.

MAIBACH, H.I. and ANJO, D.M. (1981), Percutaneous penetration of benzene and benzene contained in solvents used in the rubber industry, *Archives of Environmental Health*, **36**, 256-60.

NAS (1981) *The alkyl benzenes*. (Washington, D.C.: National Academy Press).

NIOSH (1976), *Criteria for a recommended standard... Occupational exposure to phenol* (Cincinnati, OH: National Institute for Occupational Safety and Health).

NIOSH (1976), *Revised recommendation for an occupational standard for benzene*, (Cincinnati: National Institute for Occupational Safety and Health).

NIOSH (1978) *Criteria for a recommended standard... Occupational exposure to cresol* (DHEW (NIOSH) publication No. 78-133) (Cincinnati: National Institute for Occupational Safety and Health).

PATRICK, E., MAIBACH, H.I. and BURKHALTER, A. (1985), Mechanisms of chemically induced skin irritation. I. Studies in time course, dose response, and components of inflammation in the laboratory mouse, *Toxicology and Applied Pharmacology*, **81**, 476-90.

PIOTROWSKI, J.K. (1971), Evaluation of exposure to phenol: absorption of phenol vapour in the lungs and through the skin and excretion of phenol in urine, *British Journal of Industrial Medicine*, **28**, 172-8.

RIIHIMÄKI, V. and PFÄFFLI, P. (1978), Percutaneous absorption of solvent vapors in man, *Scandinavian Journal of Work and Environmental Health*, **4**, 73-85.

RINSKY, R.A., SMITH, A.B., HORNUNG, R., FILLOON, T.G., YOUNG, R.Y., OKUM, A.H. and LANDRIGAN, P.J. (1987), Benzene and leukemia: an epidemiological risk assessment, *New England Journal of Medicine*, **316**, 1044-50.

ROBERTS, M.S., ANDERSON, R.A. and SWARBRICK, J. (1977), Permeability of human epidermis to phenolic compounds, *Journal of Pharmacy and Pharmacology*, **29**, 677-83.

SANDERS, C.L., SKINNER, C. and GELMAN, R.A. (1986), Percutaneous absorption of 7,10^{14}C-benzo[a]pyrene and 7,12^{14}C-dimethylbenz[a]anthracene in mice, *Journal of Environmental Pathology, Toxicology and Oncology*, **7**, 25-34.

SATO, A. and NAKAJIMA, T. (1978), Differences following skin or inhalation exposure in the absorption and excretion kinetics of trichloroethylene and toluene, *British Journal of Industrial Medicine*, **35**, 43-49.

SMYTH, H.F., CARPENTER, C.P., WEIL, C.S., POZZANI, U.C., STRIEGEL, J.A. and NYCUM, J.S. (1969), Range-finding toxicity data: List VII *American Industrial Hygiene Association Journal*, **30**, 470-476.

SOUTHWELL, D., BARRY, B.W. and WOODFORD, R. (1984), Variation in permeability of human skin, within and between specimens. *International Journal of Pharmacy*, **18**, 299-309.

SUSTEN, A.S., DAMES, B.L., BURG, J.R. and NIEMEIER, R.W. (1985), Percutaneous penetration of benzene in hairless mice: an estimate of dermal absorption during tire-building operations, *Americal Journal of Industrial Medicine*, **7**, 323-35.

TSURATA, H. (1982), Percutaneous absorption of organic solvents, III. On the penetration rates of hydrophobic solvents through the excised rat skin, *Industrial Health*, **20**, 335-345.

TSURATA, H., IWASAKI, K. and KANNO, S. (1987), A method for calculating the skin absorption rate from the amount retained in the whole body of skin-absorbed toluene in mice, *Industrial Health*, **25**, 215-20.

VALETTE, G. and CAVIER, R. (1954), Percutaneous absorption and chemical composition of hydrocarbons, alcohols and esters (in French), *Archives of Internationales de Pharmacodynamie*, **97**, 232-40.

VERNOT, E.H., MACEWEN, J.D., HAUN, C.C. and KINKEAD, E.R. (1977), Acute toxicity and skin corrosion data for some organic and inorganic compounds and aqueous solutions, *Toxicology and Applied Pharmacology*, **42**, 417-23.

WESTER, R.C. and MAIBACH, H.I. (1984), Advances in percutaneous absorption. In *Cutaneous Toxicity,* eds V.A. Drill and P. Lazar, pp. 29-40 (New York: Raven).

WHEELER, L.A., SAPERSTEIN, M.D. and LOWE, N.J. (1981), Mutagenicity of urine from psoriatic patients undergoing treatment with coal tar and ultra-violet light, *Journal of Investigative Dermatology,* **77**, 181-5.

WOHLRAB, W. and WOZNIAK, K.D. (1984), Epidermale Reaktionskinetik nach Benzol-Kontakt, *Dermatosen,* **32**, 211-4.

Chapter 7
Halogenated cyclic compounds

Introduction

The halogenated cyclic compounds most frequently considered to be a skin absorption hazard are indicated in table 7.1. The most important of these are considered separately below.

Halogenated aromatic compounds include some chemicals that have been widely used, but some are poorly defined. For example, most countries list the group of chlorinated naphthalenes as a skin hazard, but the FRG only has a 'skin' denotation for the pentachlorinated congeners, (ACGIH) for both penta- and hexachlorinated, and some other countries list several individual congeners. Also, polychlorinated biphenyls (PCBs) are sometimes listed as the individual chemical products (e.g. Aroclor in Japan).

Halogenated aromatic compounds with a 'skin' denotation in single countries are 1,2-dichlorobenzene, *alpha, alpha, alpha*-trifluorotoluene, *o*-chlorotoluene (2-chloro-1-methylbenzene), *p*-chlorophenol, trichlorophenol, tetrachlorophenol, 2,4,5-T (2,4,5-trichlorophenoxyacetic acid), and chlorodiphenyl oxides. Hexachlorocyclopentadiene is the only halogenated cycloalkene with a 'skin'

Table 7.1. Halogenated cyclic compounds that are considered a skin hazard in several countries (y = yes, n = no, o = other regulation). Those that are described in separate sections are italicized.

CAS No.	Chemical	Number of countries	FRG	Sweden	USA
Halogenated aromatic compounds					
1336-36-3	*PCBs*	9	y	y	y
50-29-3	*DDT*	6	y	y	y
1321-64-8	*Chloronaphthalene* (penta-)	14	y	y	y
87-86-5	*Pentachlorophenol*	15	y	y	y
Other halogenated cyclic compounds					
58-89-9	*Lindane* (*gamma*-hexachlorocyclohexane)	11	y	n	y
8001-35-2	*Toxaphene* (chlorinated camphene)	8	y	n	y
57-74-9	Chlordane	10	y	n	y
76-44-8	Heptachlor	9	y	n	y
309-00-2	Aldrin	11	y	n	y
60-57-1	Dieldrin	12	y	n	y
72-20-8	Endrin	9	y	n	y

denotation not otherwise mentioned. Many of the chlorinated cyclic compounds do not appear on current official lists, because they have been banned, or the usage has been severely restricted. Thus, the absence of 'skin' denotation in some cases does not indicate that the compound is not considered a skin hazard; rather, the compound has adverse effects serious enough to limit its use and therefore delete it from the list.

Compounds of large molecular size will tend to move very slowly through the skin, and those that are very lipophilic may perhaps penetrate through hair follicles and sweat glands. Some halogenated aromatics, such as bromobenzene, have a high solubility in psoriasis scales (Hansen and Andersen, 1988). As chronic toxicity, and in some cases carcinogenicity, are the critical effects, even a slow skin penetration can be of concern in long-term exposures. Percutaneous absorption seems to have played an important role in occupational exposures to several of these compounds (Kimbrough and Grandjean, 1990).

Polychlorinated biphenyls

Skin exposure potential

Polychlorinated biphenyls (PCBs, chlorinated diphenyls) were introduced in about 1930 as dielectric fluids in capacitors and transformers. Although PCBs during past years have been employed for several other purposes, the use has now been decreased considerably and is mainly limited to closed systems. Although current production is decreasing, a considerable skin exposure potential exists in relation to leakage of electrical equipment and hydraulic systems. Surface contamination in one capacitor production facility generally ranged from 0.4 to 6 μg cm^{-2} (Maroni et al., 1981), while chronic leaking in another facility resulted in higher levels (Elo et al., 1985). PCBs have been identified in skin wipes from capacitor workers at levels averaging 10–19 μg cm^{-2} (Maroni et al., 1981), and several other studies reviewed by Kimbrough and Grandjean (1989) document that tools and work clothes may be contaminated, thus leading to extended skin exposure.

Physicochemical properties

A total of 209 possible congeners may result from total or partial chlorination of a biphenyl, and the commercial PCB products which contain mixtures of these congeners, vary considerably in properties. Depending on the production method and subsequent exposure to heat, PCB may contain chlorinated dibenzofurans. At room temperature, a PCB product may be an oily fluid or a white solid. The solubility in water is very low, and the log P_{ow} varies from 3.8 to 11.2 (EPA, 1986). Furthermore, 1.5 mg of powdered stratum corneum from human calluses absorbed almost all of the PCB (54% chlorine) present in a 1.5 ml aqueous solution (Wester et al., 1987). The calculated solubility of trichlorobiphenyl in psoriasis scales suggests that considerable dermal uptake of this compound could occur (Hansen and Andersen, 1988).

Experimental data

Percutaneous penetration of a radio-labelled PCB product (Kanechlor-400) was documented by autoradiography, and passage via hair follicles appeared important (Nishizumi, 1976). When ^{14}C-labelled PCB was dermally applied to guinea pigs, 33.2% of PCB with 42% chlorine content was absorbed, and 55.6% of PCB with 54% chlorine; immediate washing with water and acetone removed little more than half of PCB applied to the skin, and later washing much less (Wester et al., 1985). Within 72 h after application of 2,4,5,2′,4′,5′-hexachlorobiphenyl to the skin of adult rats, a total of 0.109 μmol cm^{-2} (38 μg cm^{-2}) had penetrated the skin, i.e. about 40% of the dose; at high dosages, the relative penetration was less (Shah et al., 1987). Dermal LD_{50} values are difficult to evaluate, as acute toxicity is of little importance in relation to PCB exposure. When applied to rat skin (25 mg kg^{-1}day^{-1} for 6 days), PCB induced a three- to four-fold increase in hepatic P-450 actitivy and an increase in liver weight and microsomal protein; the effects of PCB were greater and persisted for a longer time than effects caused by dermal exposure to DDT (Bickers et al., 1973). Some PCBs, in particular those with higher chlorination, are metabolized and excreted very slowly (EPA, 1986). Thus, dermal absorption may lead to accumulation in the body. Chronic toxicity includes effects on the skin, liver and immune system, possible carcinogenicity and foetotoxicity (EPA, 1986). A large number of experimental studies have examined the provocation of chloracne as a result of cutaneous exposure, but this effect is primarly due to toxic contaminants. PCB (particularly with greater than 50% chlorination) is regarded a probable carcinogen (class 2A) (IARC, 1987).

Human data

When respiratory exposure to PCBs is low, as in transformer maintenance under controlled condition, dermal absorption appears to be the dominant pathway for PCB levels in the body (Lees et al., 1987). Data from various sources lead Wolff (1985) to conclude that exposure of the skin to about 5 μg cm^{-2} during capacitor production could contribute 20% of the total uptake during chronic exposure situations. The human health effects of PCB exposure may depend on the possible presence of various contaminants. Transient skin irritation, chloracne and liver toxicity, and possibly respiratory impairment and immunotoxicity have been reported in relation to PCB exposures (Kimborough and Grandjean, 1989). Limited evidence is available that PCBs are carcinogenic to humans, and they are considered a class 2A carcinogen (IARC, 1987).

Conclusion

Polychlorinated biphenyls (PCBs) constitute a complex group of compounds, sometimes contaminated by more trace impurities. Available data on skin absorption and systemic toxicity does not allow any differentiation between various congeners and contaminants. PCBs may penetrate the skin and significantly contribute to the total body burden from cumulated occupational exposure to these compounds. Due to chloracne and other toxic effects, including probable carcinogenicity, PCBs should be regarded as a skin exposure hazard.

Chlorinated naphthalenes

Skin exposure potential

Theoretically, 76 individual isomers of chlorinated naphthalenes are possible. The commercial products are usually mixtures with various degrees of chlorination. Various chlorinated naphthalene products have been used for several purposes in the past, but their use has declined considerably. The production in the USA in 1979 was less than 270 tonnes, with less than 5 per cent being pentachlorinated naphthalene and none of the more highly chlorinated compounds (EPA, 1980). Chlorinated naphthalenes are used as additives to engine oils and cutting oils, as capacitor dielectrics (e.g. in automobiles) and as electroplating stopoff compounds (EPA, 1980). Data on skin contact is very limited, but outbreaks of chloracne in relation to several different uses would indicate that cutaneous exposure has occurred.

Physicochemical properties

Chlorinated naphthalenes are water-insoluble waxy substances with a high chemical stability. As the chlorine content increases, the melting point increases. For the commercial mixture of penta- and hexachlorinated napthalenes, the melting point (softening point) is 137°C (EPA, 1975); for 1,2,3,4,5-pentachloro-naphthalene, the melting point is 168.5°C (EPA, 1980). With the large molecular size and lipophilic character, a protracted absorption could take place because of the low vapour pressure. These compounds dissolve readily and concentrate in the sebum of the hair follicles (Jones, 1941).

Experimental data

No recent information on experimental assessment of skin absorption is available. Application of used crank case oil containing chlorinated naphthalenes to the back of a cow (250 mg oil/week) resulted in hyperkeratosis and signs of systemic toxicity; other species appear less susceptible, but quantitative data on skin penetration are unavailable (EPA, 1980). Also, the evidence available does not allow any detailed evaluation of the toxicity of the various congeners involved. However, von Wedel et al., (1943) reported that the penta- and hexachloronaphthalenes were the most toxic congeners, and this conclusion has been supported by more recent evidence (EPA, 1980).

Human data

Although cases of chloracne have also resulted from oral intake of chlorinated naphthalenes, this characteristic lesion has been used as a specific marker for occupational exposure (mainly cutaneous) to these compounds. Only penta- and hexachloronaphthalene appear to cause chloracne, while lower chlorination is without effect (Shelley and Kligman, 1957). Chloracne also develops on skin areas not directly exposed to the compounds. All 59 workers exposed to chlorinated naphthalenes during production of electrical coils exhibited dermatological abnormalities, of which those in 56 could be attributed to the highly chlorinated naphthalenes (Kleinfeld et al., 1972). The pathogenesis of chloracne has been studied in detail (Plewig, 1970); changes in cell differentiation may be involved. Many cases of severe poisoning have occurred, and death was usually due to liver toxicity (Kimbrough and Grandjean, 1989). However, the exact contribution of skin absorption and the identity of the causative congeners have not been established. Also, detailed dose-response data are unavailable (EPA, 1980).

Conclusions

Chlorinated naphthalenes are a group of chemicals of which the penta- and hexachlorinated compounds have caused documented adverse health effects, including chloracne and liver toxicity. Although detailed information on skin penetration rates is absent, the physicochemical properties and the induction of chloracne as a result of occupational exposures would strongly suggest a skin exposure hazard.

Pentachlorophenol

Skin exposure potential

Technical pentachlorophenol (PCP) contains tetrachlorophenol, chlorinated phenoxyphenols and traces of chlorinated dibenzofurans and dibenzo-p-dioxins. PCP has important insecticidal and fungicidal actions and is used mainly as a wood preservative, but PCP, sodium pentachlorophenate and PCP fatty acid esters are also applied to protect against microbiological attacks in the paint, paper, rubber and textile industries. Production within the EC was about 800 tonnes in 1979, decreasing from 13 000 tonnes in 1974. Considerable potentials for skin exposures exist in the production and user industries.

Physicochemical properties

At room temperature, pure PCP forms colourless crystals with a phenolic odour. The vapour pressure at 20°C is about 1×10^{-4} mm Hg. The melting point for the anhydrous form is 191°C. PCP has a solubility in water of 14 mg l^{-1} at 20°C, while sodium pentachlorophenate is easily water soluble. PCP is soluble in many organic solvents, but it has a limited solubility in alkanes, carbon tetrachloride and hydrocarbon oils. The n-octanol–water partition coefficient (log P_{ow}) has been reported as 3.81–5.12, probably depending on the purity and the analytical method. On the basis of physicochemical properties, a penetration rate of 12 µg cm^{-2}h^{-1} has been calculated (Fiserova-Bergerova and Pierce, 1989).

Experimental data

LD$_{50}$ values for various cutaneous applications of PCP ranged from 40 mg kg^{-1} and upwards, and oral LD$_{50}$ were of the same magnitude; for 5% PCP in fuel oil, the cutaneous LD$_{50}$ was 60–70 mg kg^{-1}, and the oral value 70–90 mg kg^{-1} (Deichmann et al., 1942). More recently, the oral LD$_{50}$ in male and female rats was found to be 146 and 175 mg kg^{-1}, while the dermal LD$_{50}$ values were 320 and 330 mg kg^{-1}, respectively (Gaines, 1969). The importance of the vehicle was illustrated by the observation that little penetration occurred when PCP was administered on the skin in olive oil or ethyl alcohol solutions, while low LD$_{50}$ values were obtained when fuel, paraffin, furnace or pine oil were used as solvents (Deichmann et al., 1942). In these experiments, skin irritation was noted at low dosages: an initial oedema developed, but later dried out leaving a wrinkled skin area after about 1 week. Inadequate evidence is available for carcinogenicity to animals (IARC, 1987).

Human data

Following immersion of the hands in a 0.4% PCP solution for 10 min, PCP was detected in the urine at a level of 0.24 mg l^{-1} 2 days later, decreasing over the subsequent month; the patient complained of a painful sensation in the hands (Bevenue et al., 1967). Skin irritation is a common result in human exposure situations, and acute irritation usually occurs from exposure to a 10% solution while repeated contact with lower concentrations may cause irritation as well. The occurrence of chloracne has been related to PCP, but the aetiology of this skin disease may not be due to pure PCP, but rather to some contaminants (Anon, 1978). At least 24 fatalities and many milder cases of PCP poisoning have been reported in workers involved in the production of the chemical, in the wood and rubber manufacturing industries and field applications of the chemical as a fungicide or molluscicide (Anon, 1978; Truhaut et al., 1952; Gordon, 1956; Menon, 1958). Over a 10-year period, 46 cases of occupational PCP poisoning were recorded in France; one case was fatal and resulted from percutaneous exposure (Efthymiou et al., 1984). Calculations have suggested that exposure of a skin area of 360 cm² to pentachlorophenol would cause an absorption 30-fold above inhalation of the compound at the TLV level for the same time period (Fiserova-Bergerova and Pierce, 1989). High fever, loss of appetite and body weight, excessive perspiration, headache, and dyspnoea progressing into coma are prominent features of acute PCP poisoning. Liver changes, porphyria and other chronic effects are probably caused by contaminants rather than by PCP *per se*. Symptoms have occurred at blood concentrations at 4–8 mg 100 ml^{-1}, but the safe dosage range is not known (Anon, 1978). Human cases of PCP poisoning have been mostly due to dust inhalation, skin contact, or both (Anon, 1978). Intoxications have been reported in knapsack sprayers whose backs were drenched or whose bare feet were wet from walking through puddles of spray material. Also the use of a PCP preparation for laundering diapers in a hospital resulted in PCP absorption and 20 cases of poisoning in newborn infants, two of the cases being fatal (Armstrong *et al.*, 1969). The exact contribution by skin absorption in the individual cases is hard to estimate, but must have been of major significance in a large number of cases. Chlorophenols as a group are considered a class 2B carcinogen (IARC, 1987).

Conclusions

Pentachlorophenol readily passes through the skin, and cutaneous absorption has been reported as a cause of several cases of human PCP poisoning. This compound should therefore be regarded as a skin absorption hazard.

Toxaphene

Skin exposure potential

Toxaphene is a mixture of polychlorinated bicyclic terpenes, predominantly chlorinated camphenes, sometimes referred to as camphechlor. It is produced by chlorination of camphene, and the product contains 67–69% chlorine by weight. Production and use have dropped considerably, but it is still used as an insecticide on vegetables, vines and cotton. In toxaphene production workers, contamination of uncovered skin was found to be 0.03–1 mg cm^{-2} (World Health Organization, 1984).

Physicochemical properties

Toxaphene is a waxy solid with a melting range of 65–90°C. The commercial product contains a minimum of 177 chlorinated compounds. It is practically insoluble in water but soluble in common organic solvents. The octanol/water coefficient (log P_{ow}) is 2.9–3.3.

Experimental data

The dermal LD_{50} for toxaphene in rodents is about 1 g kg^{-1}, though considerably lower in rabbits when applied as a solution in peanut oil; these levels are much higher than the LD_{50} values obtained by other routes of administration (WHO, 1984). Daily dermal application of toxaphene in a kerosene solution at 100 mg kg^{-1} caused mortality within 6 days in both rabbits and guinea pigs (McGee et al., 1952). However, daily dermal application of up to 600 mg kg^{-1} to dogs for 30 days caused no apparent adverse effects, but degenerative changes in the liver were seen on autopsy (Lackey, 1949). Toxaphene causes cancer in experimental animals (IARC, 1987).

Human data

One case of toxaphene poisoning has been reported as a result of possible percutaneous absorption; a 70-year-old man had his hands in contact with a toxaphene-lindane solution for 2 h, and 8 h later, headache, nausea, and poor co-ordination developed; over the following week, the muscles became flaccid and then the patient became semicomatose (Pollock, 1958). Otherwise, systemic effects include muscle spasms, tremors, convulsions and coma (McGee et al., 1952; WHO, 1984). Several cases of aplastic anaemia have been linked to toxaphene exposures, some being dermal (WHO, 1984). Toxaphene is considered a possible carcinogen (class 2B) (IARC, 1987).

Conclusions

Available evidence suggests that toxaphene may be absorbed through the skin, in particular when skin contact is prolonged. Toxaphene should therefore be regarded a skin absorption hazard.

Other chlorinated aromatic compounds

DDT

For dichlorodiphenyltrichloroethane (DDT), the dermal LD_{50} is much higher than the value obtained by oral route (Gaines, 1969). In mice, DDT penetrated the skin more slowly than other pesticides (Shah et al., 1981). While ^{14}C-labelled DDT seems to penetrate quite effectively through the skin of rabbit and pig, much less was absorbed in humans and monkeys (Bartek and La Budde, 1975). Sufficient evidence is available for carcinogenicity to animals, and DDT is regarded as a possible carcinogen (class 2B) (IARC, 1987).

Chlorinated benzenes and phenols

For chlorobenzene, a theoretical penetration rate of 0.24 mg cm^{-2}h^{-1} has been calculated, thus suggesting that percutaneous absorption could contribute significantly to the total exposure (Fiserova-Bergerova and Pierce, 1989).

Tetrachlorophenols and trichlorophenols or their salts are, along with pentachlorophenol, extensively used as wood preservatives; the commercial products contain chlorinated dibenzofurans, diphenylethers and dibenzodioxins as contaminants. A study of urinary chlorophenol excretion in sawmill workers suggested that respiratory exposures could not explain the variations, and that percutaneous absorption was significant (Kauppinen and Lindroos, 1985).

2,4-D and 2,4,5-T

For 2,4-dichlorophenoxyacetic acid (2,4-D), percutaneous absorption was demonstrated after application of the ^{14}C-labelled compound to the forearm of human volunteers; about 6% of the dose applied was excreted in the urine within 5 days (Feldman and Maibach, 1974). The dimethylamine salt of 2,4-D in aqueous solution showed a peak skin penetration after about 30 min, but absorption continued and reached about 10% of the dose applied after 72 h (Pelletier et al., 1989). The vast majority of 2,4-D absorbed in pesticide sprayers is due to the skin contamination (Lavy et al., 1982, Libich et al., 1984, Abbot et al., 1987). The same applies to 2,4,5-trichlorophenoxyacetic acid (2,4,5-T) (Lavy et al., 1980). The hands receive the largest dose, but knapsack spraying causes a considerable potential for exposure of the lower legs, the total dermal dose often exceeding 100 mg h^{-1} (Abbott et al., 1987). Chronic peripheral neuropathy has occurred as a delayed effect of cutaneous exposure to weed-killers containing esters of 2,4-D (Goldstein et al., 1959). Limited evidence is available for carcinogenicity of chlorophenoxy herbicides in humans, and the evidence for carcinogenicity to animals is inadequate for 2,4-D and 2,4,5-T; chlorophenoxy herbicides as a group belong to the possible carcinogens (class 2B) (IARC, 1987).

Chlorinated dibenzofurans and dibenzodioxins

Experimental studies on rats suggest that up to 50% dermal absorption can occur within three days after application of small amounts of 2,3,7,8-tetrachlorodibenzofuran and 2,3,7,8-tetrachlorodibenzo-p-dioxin (TCDD), while lesser absorption was seen with two pentachlorinated dibenzofurans (Brewster et al., 1989). TCDD is considered a possible carcinogen (class 2B) (IARC, 1987).

Other compounds

Benzene sulfonyl chloride is used as an intermediate in the production of dyes, pesticides, drugs, etc.; three workers splashed in the face with this compound developed severe eye and skin irritation, one had reversible hepatomegaly, and one developed anaphylactic shock (Stasik, 1975). Individual hyper-reactivity may have played a role in theses cases.

Within 72 h after application of chlordecone (kepone) in acetone solution to the skin of adult rats, only 1% absorption was detected, i.e. the lowest fraction of 14 pesticides studied; a higher degree of absorption was seen at lower doses (Shah et al., 1987). The compound is considered a possible carcinogen (class 2B) (IARC, 1987).

In the same study, permethrin was applied in an acetone solution to the skin of adult rats, and more than half of the lowest dose was absorbed in 72 h; only about one-fifth was absorbed at higher doses (Shah et al., 1987). This chlorinated compound belongs to the biological insecticides with a low mammalian toxicity.

Other chlorinated cyclic compounds

Lindane

Lindane (*gamma*-hexachlorocyclohexane) may also be absorbed through the skin. Topical application of lindane in an acetone solution to the palms of rhesus monkeys resulted in an absorption of more than 50% within 24 h; lower absorption rates were seen after application on the forehead and the forearm (Moody and Ritter, 1989). A single topical treatment with a 1% lindane cream resulted in detectable lindane concentrations in the blood of the patient and an increase in antipyrine clearance suggesting induction of microsomal enzymes in the liver (Hosler et al., 1979). Very high serum lindane concentrations have been documented in lindane-treated scabies patients with severe skin lesions that probably increased the absorption of the compound (Lange et al., 1981).

An 18-month-old infant was treated prophylactically against scabies with a lindane lotion; several hours after the second application, the child had generalized seizures, and subsequent determination of the lindane concentration in serum suggested lindane as the cause (Telch and Jarvis, 1982).

Dermal exposure to lindane can also cause urticaria (Fisher, 1986). However, no adverse systemic effects due to percutaneous uptake of lindane have been reported from industrial exposures.

Chlordane and heptachlor

The use of chlordane as a termicide and other purposes has been discouraged or banned. Dermal application of chlordane to rats was as toxic as oral administration (Ambrose et al., 1953; Gaines, 1969). The rate of skin penetration is not known, but some chlordane may persist on the skin, as hexane-washing of the hands of a former chlordane applicator showed traces of the pesticide 2 years after last exposure (Kazen et al., 1974). One fatality has been recorded in relation to cutaneous exposure to chlordane: a woman spilled a suspension containing chlordane and DDT on the front of her clothing, and less than 1 h later she became confused and had generalized convulsions; the subsequent death was attributed to chlordane, as pathological findings concurred with those seen in experimental chlordane toxicity (Derbes et al., 1955). Limited evidence is available that chlordane is carcinogenic in experimental animals (IARC, 1987), but prudence would suggest that any remaining exposures should therefore be limited as much as possible.

With the related compound, heptachlor, less information is available. Dermal LD_{50} values would suggest that a skin hazard could be present, but the main concern relates to cumulative exposures, chronic effects and potential carcinogenicity (Reuber, 1978).

Aldrin, dieldrin and endrin

Some very toxic chlorinated compounds are not produced any more or are used for such limited purposes that skin contact is unlikely to occur. In particular, aldrin, dieldrin, and endrin all cause convulsions in acute intoxication, but may also lead to cumulative toxicity (Jager, 1970).

The dermal and oral LD_{50} values for aldrin and dieldrin are of the same order of magnitude (Gaines, 1969; WHO, 1990). The dermal toxicity to rabbits increased when aldrin and dieldrin were applied in olive oil instead of dry powder, and they were more toxic when dissolved in kerosene (WHO, 1990). The toxicity depends on the percentage of aldrin and dieldrin in the formulation, the solvent used, and the physical form of the formulation (WHO, 1990). After absorption, aldrin is metabolized to dieldrin, and the two compounds therefore share toxic potentials. Limited evidence is available for carcinogenicity of both aldrin and dieldrin to animals (IARC, 1987).

Penetration of human skin was demonstrated with ^{14}C-labelled aldrin; about 8% of the dose applied was excreted in the urine within 5 days (Feldman and Maibach, 1974). Unfortunately, in human cases of aldrin poisoning (Nelson, 1953; Kazantzis *et al.*, 1964), the evidence does not allow an evaluation of the possible significance of dermal absorption.

A review of international experience with the use of dieldrin in antimalarial programmes indicated that skin contamination is the greatest hazard under field conditions (Hayes, 1959). Percutaneous absorption was demonstrated after application of ^{14}C-labelled dieldrin to the forearm of human volunteers; as in the case of aldrin, about 8% of the dose applied was excreted in the urine within 5 days (Feldman and Maibach, 1974). Comparatively large amounts of dieldrin were absorbed by a pesticide formulator with scleroderma, thus indicating easier penetration of damaged skin (Starr and Clifford, 1971). As also seen with chlordane, hexane-washing of the hands of a former pesticide applicator showed traces of the pesticide 2 years after last exposure (Kazen *et al.*, 1974).

Endrin is a stereoisomer of dieldrin. LD_{50} values after cutaneous application in a xylene solution are somewhat similar to those after oral intake in rats (Gaines, 1969). Lower toxicity would be expected for endrin as a dry powder. Dermal exposure to endrin in orchard sprayers was estimated to be 2.5–3 mg h^{-1}, i.e. much above inhalation exposures; even higher skin exposures were seen when dusting potatoes (Wolfe *et al.*, 1963, 1967), but definite human evidence of systemic toxicity due to percutaneous uptake is lacking.

References

ABBOTT, I.M., BONSALL, J.L., CHESTER, G., HART, T.B. and TURNBULL, G.J. (1987), Worker exposure to a herbicide applied with ground sprayers in the United Kingdom, *American Industrial Hygiene Association Journal,* **48**, 165-75.

AMBROSE, A.M., CHRISTENSEN, H.E., ROBBINS, D.J. and RATHER, L.J. (1953). Toxicological and pharmacological studies on chlordane, *Archives of Industrial Hygiene and Occupational Medicine,* **7**, 197-210.

ANON (1978), Pentachlorophenol. In *Criteria (dose/effect relationships) for organochlorine pesticides* (Oxford: Pergamon), pp. 157-176.

ARMSTRONG, R.W., EICHNER, E.R., KLEIN, D.E., BARTHEL, W.F., BENNETT, J.V., JONSSON, V., BRUCE, H. and LOVELESS, L.E. (1969), Pentachlorophenol poisoning in a nursery for newborn infants II. Epidemiologic and toxicologic studies, *Journal of Pediatrics,* **75**, 317-321.

BARTEK, M.J. and LA BUDDE, J.A. (1975), Percutaneous absorption in vitro. In *Animal Models in Dermatology,* ed. H.I., Maibach (New York: Churchill Livingstone), pp. 103-20.

BEVENUE, A., HALEY, T.J. and KLEMMER, H.W. (1967). A note on the effects of a temporary exposure of an individual to pentachlorophenol, *Bulletin of Environmental Contamination and Toxicology,* **2**, 293-296.

BICKERS, D.R., KAPPAS, A. and ALVARES, A.P. (1973), Effects of chlorinated hydrocarbons applied to skin on hepatic cytochrome P-450 and drug metabolizing enzymes (Abstract), *Federation Proceedings,* **32**, 235.

BREWSTER, D.W., BANKS, Y.B. CLARK, A.M. and BIRNBAUM, L.S. (1989), Comparative dermal absorption of 2,3,7,8-tetrachlorodibenzo-*p*-dioxin and three polychlorinated dibenzofurans, *Toxicology and Applied Pharmacology,* **97**, 156-66.

DEICHMANN, W., MACHLE, W., KITZMILLER, K.V. and THOMAS, G. (1942), Acute and chronic effects of pentachloropehnol and sodium pentachlorophenate upon experimental animals, *Journal of Pharmacology and Experimental Therapy,* **76**, 104-117.

DERBES, V.J., DENT, J.H., FORREST, W.W. and JOHNSON, M.F. (1955), Fatal chlordane poisoning, *Journal of the American Medical Association,* **158**, 1367-9.

EFTHYMIOU, M.L., CONSO, F. and FOURNIER, E. (1984), Intoxication professionelle par le pentachlorophenol et ses analogues — a propos de 63 cas. XXI International Congress on Occupational Health, Abstracts, Dublin, p. 11.

ELO, O., VUOJOLAHTI, P., JANHUNEN, H. and RANTANEN, J. (1985), Recent PCB accidents in Finland, *Environmental Health Perspective,* **60**, 315-319.

EPA (1975), *Environmental hazard assessment report, Chlorinated naphthalenes,* (EPA-560/001) (Washington: D.C.: U.S. Environmental Protection Agency).

EPA (1980), Ambient water quality criteria for chlorinated naphthalene, (EPA-440/5-80-031) (Washington, D.C.: U.S. Environmental Protection Agency).

EPA (1986), *Drinking water criteria document for polychlorinated biphenyls (PCBs),* (Final draft EPA-600/x-84-198) (Cincinnati, Ohio: U.S. Environmental Protection Agency).

FELDMAN, R.J. and MAIBACH, H.I. (1974), Percutaneous penetration of some pesticides and herbicides in man, *Toxicology and Applied Pharmacology,* **28**, 126-32.

FISEROVA-BERGEROVA, V. and PIERCE, J.T. (1989), Biological monitoring V. Dermal absorption, *Applied Industrial Hygiene,* **4**, F14-F21.

FISHER, A.A. (1986), *Contact dermatitis,* 3rd edn. (Philadelphia, Lea & Febiger).

GAINES, T.B. (1969), Acute toxicity of pesticides, *Toxicology and Applied Pharmacology,* **14**, 515-34.

GOLDSTEIN, N.P., JONES, P.H. and BROWN, J.R. (1959), Peripheral neuropathy after exposure to an ester of dichlorophenoxyacetic acid, *Journal of the American Medical Association*, **171**, 1306-9.
GORDON, D. (1956), How dangerous is pentachlorophenol? *Medical Journal of Australia*, **2**, 485-488.
HANSEN, C.M. and ANDERSEN, B.H. (1988), The affinities of organic solvents in biological systems, *American Industrial Hygiene Association Journal*, **49**, 301-8.
HAYES, W.J. (1959), The toxicity of dieldrin to man, *Bulletin of the World Health Organisation*, **20**, 891-912.
HOSLER, J., TSCHÄNZ, C., HIGNITE, C.E. and AZARNOFF, D.L. (1979), Topical application of lindane cream (Kwell) and antipyrine metabolism, *Journal of Investigative Dermatology*, **74**, 51-3.
IARC (1987), *Overall evaluations of carcinogenicity: An updating of IARC Monographs Volumes 1 to 42*, (IARC Monographs on the evaluation of carcinogenic risk of chemicals to man Suppl. 7). (Lyon: International Agency for Research on Cancer).
JAGER, K.W. (1970), *Aldrin, dieldrin, endrin and telodrin*. (Amsterdam: Elsevier).
JONES, A.T. (1941), The etiology of acne with special reference to acne of occupational origin, *Journal of Industrial Hygiene and Toxicology*, **23**, 290-312.
KAUPPINEN, T. and LINDROOS, L. (1985), Chlorophenol exposure in sawmills, *American Industrial Hygiene Association Journal*, **46**, 34-8.
KAZANTZIS, G., MCLAUGHLIN, A.I.G. and PRIOR, P.F. (1964), Poisoning in industrial workers by the insecticide aldrin, *British Journal of Industrial Medicine*, **21**, 46-51.
KAZEN, C., BLOOMER, A., WELCH, R. OUDBIER, A. and PRICE, H. (1974), Persistence of pesticides on the hands of some occupationally exposed people, *Archives of Environmental Health*, **29**, 315-8.
KIMBROUGH, R.D. and GRANDJEAN, P. (1989) Occupational exposure. In *Halogenated biphenyls terphenyls, naphthalenes, dibenzodioxins,* eds R.D. Kimbrough and A.A. Jensen, pp. 485-507 (Amsterdam: Elsevier).
KLEINFELD, M., MESSITE, J. and SWENCICKI, R. (1972), Clinical effects of chlorinated naphthalene exposure. *Journal of Occupational Medicine*, **14**, 377-9.
LACKEY, R.W. (1949). Observations on the percutaneous absorption of toxaphene in the rabbit and dog. II. *Journal of Industrial Hygiene and Toxicology*, **31**, 155-7.
LANGE, M., NITZSCHE, K. and ZESCH, A. (1981), Percutaneous absorption of lindane in healthy volunteers and scabies patients, *Archives of Dermatological Research*, **271**, 387-99.
LAVY, T.L., SHEPARD, J.S., and MATTICE, J.D., (1980), Exposure measurements of applicators spraying 2,4,5-trichlorophenoxy acetic acid in the forest, *Journal of Agricultural Food Chemistry*, **28**, 626-30.
LAVY, T.L., WALSTAD, J.D., FLYNN, R.R. and MATTICE, J.D. (1982), 2,4-D exposure received by aerial application crews during forest spray operations, *Journal of Agricultural Food Chem*, **30**, 375-81.
LEES, P.S.J., CORN, M. and BREYSSE, P.N. (1987), Evidence for dermal absorption as the major route of body entry during exposure of transformer maintenance and repairmen to PCBs, *American Industrial Hygiene Association Journal*, **48**, 257-64.
LIBICH, S., TO, J., FRANK, R. and SIRONS, G. (1984), Occupational exposure of herbicide applicators to herbicides used along electric power transmission line right-of-ways, *American Industrial Hygiene Association Journal*, **45**, 52-62.
MARONI, M., COLOMBI, A., CANTONI, S., FERIOLO, E. and FOA, V. (1981), Occupational exposure to polychlorinated biphenyls in electrical workers. I. Environmental and blood polychlorinated biphenyls concentrations, *British Journal of Industrial Medicine*, **38**, 49-54.

McGee, L.C., Reed, H.L. and Fleming, J.P., (1952), Accidental poisoning by toxaphene, review of toxicology and case reports, *Journal of the American Medical Association,* **149,** 1124–6.

Menon, J.A., (1958), Tropical hazards associated with the use of pentachlorophenol, *British Medical Journal,* **1,** 1156–1158.

Moody, R.P. and Ritter, L. (1989), Dermal absorption of the insecticide lindane (1δ,2δ,3β,4δ,5δ,6β-hexachlorocyclohexane) in rats and rhesus monkeys: effect of anatomical site, *Journal of Toxicology and Environmental Health,* **28,** 161–9.

Nelson, E. (1953), Aldrin poisoning, *Rocky Mountain Medical Journal,* **50,** 483–6.

Nishizumi, M. (1976) Radioautographic evidence for absorption of polychlorinated biphenyls through the skin, *Industrial Health,* **14,** 41–4.

Pelletier, O., Ritter, L., Caron, J. and Somers, D. (1989), Disposition of 2,4-dichlorophenoxyacetic acid dimethylamine salt by Fisher 344 rats dosed orally and dermally, *Journal of Toxicology and Environmental Health,* **28,** 221–34.

Plewig, G. (1970), Zur Kinetik der Comedone-Bilddung bei Chloracne (Halowaxacne), *Archiv für Klinische und Experimentelle Dermatologie,* **238,** 228–41.

Pollock, R.W. (1958), Toxaphene-lindane poisoning by cutaneous absorption, *Northwest Medicine,* **57,** 325–6.

Reuber, M.D. (1978), Carcinomas and other lesions of the liver in mice ingesting organochlorine pesticides, *Clinical Toxicology,* **13,** 231–56.

Shah, P.V., Monroe, R.J. and Guthrie, F.E., (1981), Comparative rates of dermal penetration of insecticides in mice, *Toxicology and Applied Pharmacology,* **59,** 414–23.

Shah, P.V., Fisher, H.L., Sumler, M.R., Monroe, R.J., Chernoff, N. and Hall, L.L. (1987), Comparison of the penetration of 14 pesticides through the skin of young and adult rats, *Journal of Toxicology and Environmental Health,* **21,** 353–66.

Shelley, W.B. and Kligman, A.M. (1957), The experimental production of acne by penta- and hexa-chloronaphthalene. *Archives of Dermatology,* **75,** 689–95.

Starr, H.G. Jr. and Clifford, N.J. (1971), Absorption of pesticides in a chronic skin disease, *Archives of Environmental Health,* **22,** 396–400.

Stasik, M.J. (1975), Klinische Erfahrung mit Benzolsulfonsäurechlorid, *Archives of Toxicology,* **33,** 123–7.

Telch, J. and Jarvis, D.A. (1982), Acute intoxication with lindane (*gamma*-benzene hexachloride), *Canadian Medical Association Journal,* **126,** 662–3.

Truhaut, R., L'Epee, P. and Boussemart, E. (1952), Recherches sur la toxicologie du pentachlorophenol II. Intoxications professionelles dans l'industrie du bois. Observations de duex cas mortels. *Archives des maladies professionnelles de médicine du travail et de sécurite sociale,* **13,** 576–577.

von Wedel, H., Holla, W.A. and Denton, J. (1943), Observations on the toxic effects resulting from exposure to chlorinated naphthalenes and chlorinated phenyls with suggestions for prevention, *Rubber Age,* **53,** 419–26.

Wester, R.C., Bucks, D.A.W., Maibach, H.I. and Anderson, J.H. (1985), Polychlorinated biphenyls: Dermal absorption, systemic elimination, and dermal wash efficiency. In *Percutaneous absorption,* eds R.L. Bronaugh and H.I. Maibach, (New York: Marcel Dekker), pp. 363–72.

Wester, R.C., Mobayen, M. and Maibach, H.I. (1987), In vivo and in vitro absorption and binding to powdered stratum corneum as methods to evaluate skin absorption of environmental chemical contaminants from ground and surface water, *Journal of Toxicology and Environmental Health,* **21,** 367–74.

WHO (1984), Camphechlor (Environmental Health Criteria 45). Geneva: World Health Organization.

WHO (1990), Aldrin and dieldrin. (Environmental health criteria 91). Geneva: World Health Organization (in press).

WOLFE, H.R., DURHAM, W.F. and ARMSTRONG, J.F. (1963), Health hazard of the pesticides endrin and dieldrin: Hazards in some agricultural uses in the Pacific Northwest, *Archives of Environmental Health,* **6,** 458–64.
WOLFE, H.R., DURHAM, W.F. and ARMSTRONG, J.F. (1967), Exposure of workers to pesticides, *Archives of Environmental Health,* **14,** 622–23.
WOLFF, M.S. (1985), Occupational exposure to polychlorinated biphenyls (PCBs), *Environmental Health Perspective,* **60,** 133–8.

Chapter 8
Isocyclic amines

Introduction

The isocyclic amines most frequently considered to be a skin absorption hazard are indicated in table 8.1. The most important of these are considered separately below. Several of these amines are used, or have been used, in the production of dyes and pharmaceuticals, and as antioxidants. A common toxic effect of most of these compounds is the production of methaemoglobinaemia, but some may cause hepatotoxicity.

Some of these compounds are carcinogenic. Thus, 2-naphthylamine (beta-naphthylamine) is a confirmed carcinogen (class 1) (IARC, 1987), and it is regarded a skin absorption hazard in some countries. In addition to those shown in the table, other isocyclic amines with a 'skin' denotation in a few countries are cyclohexylamine, N-isopropylaniline (N-(2-propyl)aniline), 4-diphenylamine, and 1-naphthylamine.

Several other aromatic nitrogen compounds are used as hair dyes. Many of them cause allergic contact dermatitis, and cross-reactions occur between compounds with an amino group in the *para*-position (Fisher, 1986). The toxic potential of these compounds is less clear. These compounds bind to the skin, and only small amounts penetrate and enter the systemic circulation.

Some chlorinated amines have also received a 'skin' denotation in some countries. This applies to 4,4'-methylenebis(2-chloroaniline) (MOCA) which is a

Table 8.1. Isocyclic amines that are considered a skin hazard in several countries (y = yes, n = no, o = other regulation). Those that are described in separate sections are italicized.

CAS No.	Chemical	Number of countries	FRG	Sweden	USA
62-53-3	*Aniline*	16	y	y	y
100-01-6	*p-Nitroaniline*	12	y	n	y
95-53-4	*o-Toluidine*	10	y	n	y
1300-73-8	*Xylidines*	12	y	n	y
29191-52-4	Anisidines (*o*- and *p*-isomers)	8	y	n	y
100-61-8	N-Methylaniline	9	y	n	y
121-69-7	N,N-Dimethylaniline	12	y	n	y
106-50-3	*p-Phenylenediamine (1,4-benzenediamine)*	11	y	y	y
92-87-5	*Benzidine (bianiline)*	6	y	o	y

113

probable carcinogen (IARC, 1987). Other compounds considered to be a skin hazard in certain countries are 3,3'-dichlorobenzidine, 4-chloro-o-toluidine (4-chloro-2-methylaniline), and 4,4'-diaminodiphenylmethane (bis(4-aminophenyl)methane, MDA); these compounds are all possible carcinogens (class 2B) (IARC, 1987). o-Toluidine (o-dimethylbenzidine) also belongs to this group and should be considered a possible carcinogen as well. Additional compounds considered to be a skin hazard in single countries are 3,4-dichloroaniline, chloroanilines (all), and o-chlorobenzylidene malononitrile (2-(2-chlorophenylmethylidene) propanedinitrile).

Aniline

Skin exposure potential

Aniline (phenylamine) may be produced by catalytic hydrogenation of nitrobenzene or amination of phenol. The aniline production within EC countries was 370 000 tonnes in 1984. Aniline is mainly used as a chemical intermediate; over 70% of the consumption in Europe is employed in the production of (4,4'-methylenediphenyl di-isocyanate (MDI)), while the remainder is used for other products, such as synthetic dyestuffs. Contamination of the skin at an average of 2 mg cm^{-2} has been documented in a study of workers with occupational exposure to aniline (Baranowska-Dutkiewicz, 1982).

Physicochemical properties

Aniline is an oily liquid with a boiling point of 184°C and a vapour pressure of 7 mm Hg at 20°C. It is moderately soluble in water (35 g l^{-1} at 25°C) and miscible with most organic solvents. One gram of aniline dissolves 29 ml of water at 20°C. The octanol/water partition coefficient (log P_{ow}) is about 0.93. The measured solubility in psoriasis scales suggests a considerable dermal uptake (Hansen and Andersen, 1988). A penetration rate of 0.64 mg cm^{-2}h^{-1} was calculated on the basis of physicochemical properties (Fiserova-Bergerova and Pierce, 1989).

Experimental data

The dermal LD$_{50}$ in rats is similar to the value obtained by oral administration. Dermal LD$_{50}$ on abraded skin for rabbits is 0.82 ml kg^{-1} and for guinea pigs 2.15 ml kg^{-1} on abraded skin and 1.29 ml kg^{-1} on intact skin (Roudabush et al., 1965). Experimental studies have documented that aniline may be easily absorbed through the skin (Izmerov, 1984). The permeability constant k_p for application on human skin in vitro is 6.5 × 10^4 cm^{-1} h^{-1}, i.e. close to the k_p for water; when applied in an ethanol:water (1:1) solution, k_p increased 10-fold (Dugard and Scott, 1984). Aniline is a moderate skin irritant. Methaemoglobinaemia is well documented, but other toxicity not related to this effect has not been totally ruled out. Limited evidence is available for the carcinogenicity of aniline hydrochloride in experimental animals (IARC, 1987).

Human data

Numerous aniline intoxications have occurred due to dermal exposure (Hamilton, 1919). Application of 0.25 ml aniline on the skin of the forearm under occlusion resulted in an absorption rate of 0.18–0.72 mg cm^{-2} h^{-1} as determined from the amount remaining under the occlusion after 5 h; the absorption rate increased with higher skin temperatures due to the increased moisture (Piotrowski, 1957). Immersion of the hands in 1% and 2% solutions of aniline resulted in a percutaneous absorption rate of 0.32–1.22 mg cm^{-2} h^{-1}, as calculated from the excretion of p-aminophenol; application of neat aniline and aniline containing 3% water on the skin of the forearm resulted in similar absorption rates of 2.5–3.0 mg cm^{-2} h^{-1} (Baranowska-Dutkiewicz, 1982). The maximum excretion of p-aminophenol occurred 4–6 h after the start of exposure (Baranowska-Dutkiewicz, 1982). That an aqueous solution resulted in almost the same percutaneous absorption as did neat aniline is probably due to the better solubility of aniline in the stratum corneum of the skin as compared to the solubility in water. Also, the absorption rate for aniline was found to increase with the humidity of the skin (Piotrowski, 1972). Immersion of both hands in a 2% aqueous solution of aniline for 30 min can result in a percutaneous absorption of about 240 mg of aniline (Baranowska-Dutkiewicz, 1982); for comparison, a respiratory exposure to 5 mg m^{-3} for 8 h will cause an uptake of less than 100 mg. Percutaneous absorption of aniline vapours has also been demonstrated (Dutkiewicz and Piotrowski, 1961). Seventeen cases of aniline poisoning were seen in babies due to percutaneous absorption of dye from freshly stamped diapers (Fairhall, 1957). Poisoning due to percutaneous uptake may result from drenched clothing (Kusters, 1982). Aniline is the most frequent cause of industrial cyanosis, and skin absorption is of main importance, in particular during hot summers when protective clothing may not be worn (Sekimpi and Jones, 1986). Methaemoglobinaemia is the most prominent sign of aniline toxicity; no symptoms are generally seen at levels below 20%, higher concentrations cause headache, dyspnoea and tachycardia, and levels above 60% are life-threatening (Harrison, 1977). Inadequate evidence is available concerning possible carcinogenicity (IARC, 1987).

Conclusions

Experimental studies and industrial experience document that skin exposure to aniline may result in considerable absorption with possible systemic effects. Aniline should therefore be regarded as a skin absorption hazard.

p-Nitroaniline

Skin exposure potential

p-Nitroaniline (4-nitroaniline) is mainly used as a corrosion inhibitor, but it is also employed as an intermediate mainly in the production of dyes, but also for pesticides, p-phenylenediamine, antioxidants and petrol additives. The annual production and consumption in EC countries in 1984 was estimated at 3–6 thousand tonnes.

Physicochemical properties

p-Nitroaniline is a solid with a melting point of 148°C. It is sparingly soluble in water (0.8 g l^{-1}), but readily soluble in ethanol and diethyl ether. The octanol/water partition coefficient (log P_{ow}) is 1.39. When 1.5 mg of powdered stratum corneum from human calluses was added to a 1.5 ml aqueous solution of this compound, 2.5% of the amount was absorbed by the stratum corneum (Wester et al., 1987). p-Nitroaniline would be expected to be absorbed through the skin, in particular if applied in hydrophilic solvents.

Experimental data

p-Nitroaniline in acetone solution was applied to shaved abdominal skin of rhesus monkeys; excretion levels in urine indicated a 100% percutaneous uptake (Bronaugh and Maibach, 1985). Penetration through excised human skin was measured by in vitro diffusion cell technique; the greatest penetration occurred during the first 2 h after exposure, but absorption was not complete (Bronaugh and Maibach, 1985). p-Nitroaniline is a powerful methaemoglobin-former which may also cause haemolysis in serious cases.

Human data

p-Nitroaniline is known to be readily absorbed through the skin (ACGIH, 1986), and rapid absorption through the intact skin is frequently the main route of entry (NIOSH, 1978). However, quantitative data are lacking. Exposure to this compound has resulted in cyanosis, headache, weakness and dyspnoea due to the formation of methaemoglobin. Nitroaniline is not an unusual cause of industrial cyanosis, and skin absorption is of main importance, in particular during hot summers when protective clothing may not be worn (Sekimpi and Jones, 1986). Although methaemoglobinaemia is the main effect, a worker with pre-existing liver disease became jaundiced after exposure to p-nitroaniline and died (Anderson, 1946). Also, one case of peripheral neuropathy developed with a delay of 3 months after an acute intoxication (Baldi and Raule, 1954).

Conclusions

Little scientific evidence is available on *p*-nitroaniline, but experimental data and industrial experience indicate that this compound may readily penetrate the skin and cause systemic toxicity. *p*-Nitroaniline should therefore be regarded a skin absorption hazard.

Other compounds

Benzonitrile and *o*-phthalodinitrile have caused human intoxications that were apparently due to extensive dermal exposures at industrial accidents (Zeller *et al.*, 1969).

o-Toluidine is a primary irritant; although the dermal LD_{50} is relatively high, skin contact should be avoided, as this compound is regarded a potential carcinogen (class 2B) (IARC, 1987).

Less information is available on xylidines, although dermal penetration has been demonstrated in rabbits (Treon *et al.*, 1949).

N-Methylaniline was more toxic than xylidine; the dermal LD_{50} for *N*-methylaniline was less than one gram and similar to the value obtained by the oral route (Treon *et al.*, 1949). A recent study in mice suggests that nitrosamines can be formed in the skin as a result of a direct reaction between *N*-methylaniline and nitrogen oxide (Mirvish *et al.*, 1988). This evidence may indicate a potential carcinogenic hazard in association with cutaneous exposures.

N,N-Dimethylaniline is a skin irritant. A dermal LD_{50} of 1.77 ml kg^{-1} in rabbits (Smyth *et al.*, 1962) suggests that this compound will rarely constitute a skin absorption hazard.

With hair dye products containing radioactively labelled compounds, 2,4-diaminoanisole, 4-amino-2-hydroxytoluene, 2-nitro-*p*-phenylenediamine, HC Blue No. 1, *p*-phenylenediamine and 2-nitro-4-aminophenol penetrated human scalp (listed in increasing order); but the flux was only about 10^{-10} mol cm^{-2}h^{-1}, and the amounts recovered in the urine were well under one percent (Wolfram, 1985). Considerable penetration was seen when 2-nitro-4-aminophenol and 2-nitro-*p*-phenylenediamine were applied in an acetone solution; about 45 and 22 per cent penetrated human skin *in vitro*, respectively, and somewhat higher absorption rates were seen in monkeys, both *in vivo* and *in vitro* (Bronaugh and Maibach, 1985). For *p*-phenylenediamine, the permeability constant was 2.4×10^{-4} cm h^{-1} when applied in aqueous solution on human skin *in vitro*, but some other hair dyes with a higher octanol/water partition coefficient showed higher permeability constants (Bronaugh and Congdon, 1984). Occlusion increased, and hydrogen peroxide decreased, percutaneous absorption of *p*-phenylenediamine in dogs (Kiese *et al.*, 1968). *In vivo* penetration of labelled *p*-toluenediamine through the skin of rats and dogs has also been demonstrated, but the total amount excreted corresponded only to 0.1–0.2% of the dose (Hruby, 1977). *p*-Phenylenediamine is a very common cause of allergic contact dermatitis and may also occasionally cause contact urticaria (Edwards and Edwards, 1984). Systemic toxicity is less well documented, but *p*-toluenediamine has been associated with aplastic anaemia (Hopkins and Manoharan, 1985).

Benzidine and benzidine-derived dyes are known to be carcinogenic (IARC, 1987), and their use has been restricted. Radioactively labelled benzidine, 3,3'-dichlorobenzidine, and 3,3'-dimethoxybenzidine were applied in acetone to the shaved skin of rats; within 24 h, about half of the amount applied of the two former compounds had penetrated the skin, but only 29% of the latter (Shah and Guthrie, 1983). Some benzidine-derived dyes may be metabolized on or in the skin of rodents, thus possibly resulting in systemic exposure to benzidine (Aldrich *et al.*, 1986).

4,4'-Diaminodiphenylmethane(MOCA) is a persistent curing agent used in production of certain plastics and is rated a class 2A carcinogen (IARC, 1987). Percutaneous absorption was demonstrated *in vitro* using human skin and MOCA in ethanol but, due to the insolubility in water, negligible amounts entered the aqueous medium on the inner side of the skin (Chin *et al.*, 1983). Although this compound seems to penetrate the skin, no quantitative estimates can be made from the *in vitro* observations.

4,4'-Diaminodiphenylmethane (methylenedianiline, MDA) is used as an epoxy resin hardener; liver toxicity was documented in several workers, with extensive cutaneous exposure being the apparent cause (McGill and Motto, 1974). It is a class 2B carcinogen (IARC, 1987).

References

ACGIH (1986), *Documentation for the threshold limit values and biological exposure indices*, 5th edn (Cincinnati: American Conference of Governmental Industrial Hygienists).

ALDRICH, F.D., BUSBY, W.F. JR and FOX, J.G. (1986), Excretion of radioactivity from rats and rabbits following cutaneous application of two ^{14}C-labelled azo dyes, *Journal of Toxicology and Environmental Health*, **18**, 347–55.

ANDERSEN, A. (1946), Acute paranitraniline poisoning, *British Journal of Industrial Medicine*, **3**, 243–4.

BALDI, G. and RAULE, A. (1954), Intossicazione professionale da paranitroanilina. Un caso clinico a sintomatologia neuritica, *Medicina de Lavoro*, **45**, 584–9.

BARANOWSKA-DUTKIEWICZ, B. (1982), Skin absorption of aniline from aqueous solutions in man, *Toxicology Letters*, **10**, 367–72.

BRONAUGH, R.L. and CONGDON, E.R. (1984), Percutaneous absorption of hair dyes: Correlation with partition coefficients, *Journal of Investigative Dermatology*, **83**, 124–7.

BRONAUGH, R.L. and MAIBACH, H.I. (1985), Percutaneous absorption of nitroaromatic compounds: in vivo and in vitro studies in the human and monkey, *Journal of Investigative Dermatology*, **84**, 180–3.

CHIN, B., TOBES, M.C. and HAN, S.S. (1983), Absorption of 4,4'-methylenebis(2-chloroaniline) by human skin, *Environmental Research*, **32**, 167–78.

DUGARD, P.H. and SCOTT, R.C. (1984), Absorption through the skin. In *The Chemotherapy of Psoriasis*, (International Encyclopedia of Pharmacology and Therapeutics, Section 110), ed. H.T.P. Baden (Oxford: Pergamon).

DUTKIEWICZ, T. and PIOTROWSKI, J. (1961), Experimental investigations on the quantitative estimation of aniline absorption in man, *Pure and Applied Chemistry,* **3,** 319-23.
EDWARDS, E.K. JR and EDWARDS, E.K. (1984), Contact urticaria and allergic contact dermatitis caused by paraphenylenediamine, *Cutis,* **34,** 87-8.
FAIRHALL, L.T. (1957), *Industrial toxicology,* p. 160. (Baltimore, MD: Williams & Wilkins).
FISEROVA-BERGEROVA, V. and PIERCE, J.T. (1989), Biological monitoring V. Dermal absorption, *Applied Industrial Hygiene,* **4,** F14-F21.
FISHER, A.A. (1986), *Contact dermatitis,* 3rd edn. (Philadelphia, Lea & Febiger).
HAMILTON, A. (1919), Industrial poisoning by compounds of the aromatic series, *Journal of Industrial Hygiene,* **1,** 200-12.
HANSEN, C.M. and ANDERSEN, B.H. (1988), The affinities of organic solvents in biological systems, *American Industrial Hygiene Association Journal,* **49,** 301-8.
HARRISON, M.R. (1977), Toxic methaemoglobinaemia, *Anaesthesia,* **32,** 270-2.
HOPKINS, J.E. and MANOHARAN, A. (1985), Severe aplastic anemia following the use of hair dye: report of two cases and review of literature, *Postgraduate Medical Journal,* **61,** 1003-5.
HRUBY, R. (1977), The absorption of *p*-toluenediamine by the skin of rats and dogs, *Food and Cosmetics Toxicology,* **15,** 595-9.
IARC (1987) *Overall evaluations of carcinogenicity: An updating of IARC Monographs Volumes 1 to 42,* (IARC Monographs on the evaluation of carcinogenic risk of chemicals to man (Suppl. 7.) (Lyon: International Agency for Research on Cancer).
IZMEROV, N.F., ed. (1984), Aniline (IRPTC Scientific reviews of Soviet literature on toxicity and hazards of chemicals 53). (Moscow: Centre of International Projects, GKNT).
KIESE, M., RACHOR, M. and RAUSCHER, E. (1968), The absorption of some phenylenediamines through the skin of dogs, *Toxicology and Applied Pharmacology,* **12,** 495-507.
KUSTERS, E. (1982), Acute toxiciteit van aniline, *Tijdschrift voor Sociale Geneeskunde,* **60,** 326-9.
MCGILL, D.B. and MOTTO, J.D., (1974). An industrial outbreak of toxic hepatitis due to methylenedianiline, *New England Journal of Medicine,* **291,** 278-82.
MIRVISH, S.S., RAMM, M.D., SAMS, J.P. and BABCOCK, D.M. (1988), Nitrosamine formation from amines applied to the skin of mice after and before exposure to nitrogen dioxide, *Cancer Research,* **48,** 1095-9.
NIOSH (1978), *Occupational health guideline for p-nitroaniline* (Cincinnati: National Institute for Occupational Safety and Health).
PIOTROWSKI, J. (1957). Quantitative estimation of aniline absorption through the skin in man, *Journal of Hygiene Epidemiology Microbiology and Immunology,* **1,** 23-32.
PIOTROWSKI, J.K. (1972), Certain problems of exposure tests for aromatic compounds, *Pracovni Lekarstvi,* **24,** 94-7.
ROUDABUSH, R.L., TERHAAR, C.J., FASSETT, D.W. and DZIUBA, S.P. (1965), Comparative acute effects of some chemicals on the skin of rabbits and guinea pigs, *Toxicology and Applied Pharmacology,* **7,** 559-65.
SEKIMPI, D.K. and JONES, R.D. (1986), Notifications of industrial chemical cyanosis poisoning in the United Kingdom 1961-80, *British Journal of Industrial Medicine,* **43,** 272-9.
SHAH, P.V. and GUTHRIE, F.E. (1983), Dermal absorption of benzidine derivated in rats, *Bulletin of Environmental Contamination and Toxicology,* **31,** 73-8.
SMYTH, H.F., CARPENTER, C.P., WEIL, C.S., POZZANI, U.C. and STRIEGEL, J.A. (1962), Range-finding toxicity data, list VI, *American Industrial Hygiene Association Journal,* **28,** 95-107.

TREON, J.F., DEICHMAN, W.B., SIGMON, H.E., WRIGHT, H. WITHERUP, S.O., HEYROTH, F.F., KITZMILLER, K.V., and KEENAN, C. (1949), The toxic properties of xylidine and monomethylaniline I. The comparative toxicity of xylidine and monomethylaniline when administered orally or intravenously to animals or applied on their skin, *Journal of Industrial Hygiene and Toxicology,* **31,** 1-20.

WESTER, R.C., MOBAYEN, M. and MAIBACH, H.I. (1987), In vivo and in vitro absorption and binding to powdered stratum corneum as methods to evaluate skin absorption of environmental chemical contaminants from ground and surface water, *Journal of Toxicology and Environmental Health,* **21,** 367-74.

WOLFRAM, L.J. (1985), Hair dye penetration in monkey and man. In *Percutaneous absorption.* eds. R. L. Bronaugh, and H.I. Maibach, pp. 409-22. (New York: Marcel Dekker).

ZELLER, H., HOFMANN, H.T., THIESS, A.M. and HEY, W. (1969), Zur Toxicität der Nitrile. *Zentralblatt für Arbeitsmedizin und Arbeitsschutz,* **19,** 225-38.

Chapter 9
Organic nitro compounds and nitrates

Introduction

The isocyclic amines most frequently considered to be skin absorption hazards are indicated in table 9.1. The most important of these are considered separately below. Some of these compounds cause methaemoglobinaemia, some cause liver toxicity or yellow discolouration of the skin.

With some of the organic nitro compounds, isomers are listed if known to be the cause of adverse effects. For example, some countries list all isomers of dinitrobenzene, while others only give the 1,3-isomer a 'skin' denotation. Industrial products may be mixed, and the decision whether or not to label individual isomers is difficult.

Other organic nitro compounds and nitrates with a 'skin' denotation in a few countries are propylene glycol 1,2-dinitrate (1,2-propanedioldinitrate), cyclonite (cyclotrimethylenetrinitramine), 2,4-dinitrophenol and 4-nitrobiphenyl.

Table 9.1. Organic nitro compounds and nitrates that are considered a skin hazard in several countries (y = yes, n = no, o = other regulation). Those that are described in separate sections are italicized.

CAS No.	Chemical	Number of countries	FRG	Sweden	USA
Aliphatic nitrates					
628-96-6	*Ethylene glycol dinitrate*	13	y	y	y
55-63-0	*Nitroglycerin*	14	y	y	y
Isocyclic nitro compounds					
479-45-8	*Tetryl* (2,4,6-trinitrophenylmethylnitramine)	11	y	n	y
Aromatic nitro compounds					
98-95-3	*Nitrobenzene*	15	y	y	y
100-00-5	*p*-Nitrochlorobenzene (4-chloronitrobenzene)	12	y	n	y
99-65-0	1,3-Dinitrobenzene	14	y	y	y
1321-12-6	Nitrotoluene	11	y	n	y
25321-14-6	Dinitrotoluene	10	y	n	y
118-96-7	*2,4,6-Trinitrotoluene*	12	y	n	y
88-89-1	*Picric acid* (2,4,6-trinitrophenol)	12	y	n	y
534-52-1	*DNOC* (4,6-dinitro-*o*-cresol)	12	y	n	y

Nitroglycol

Skin exposure potential

Nitroglycol (1,2-ethanedioldinitrate, ethylene glycol dinitrate, EGDN) is highly explosive; it lowers the melting point of nitroglycerin, thus reducing the hazard associated with the use of frozen dynamite. Skin exposure to this compound is a common problem in dynamite production.

Physicochemical properties

Nitroglycol is a colourless, odourless liquid at room temperature and has a vapour pressure of 0.043 mm Hg. It is almost insoluble in water, but is soluble in ethyl alcohol, ethyl ether and benzene. The molecular weight is 152.

Experimental data

Little experimental evidence on percutaneous absorption is available, although the haemodynamic effects of this compound are well documented.

Human data

Nitroglycol could be detected in the blood within 10 min of the onset of skin exposure in a human volunteer (Hogstedt and Ståhl, 1980). In a simulated exposure experiment where a subject wore special gloves and handled a small piece of dynamite, very high nitroglycol levels were detected in the blood, and the levels differed considerably between the two arms (Hogstedt and Ståhl, 1980). Application of 100 mg of explosive material containing about 22% nitroglycol on the forearms of 6 individuals resulted in an absorption of about 3 mg cm^{-2} during a period of 7 h, i.e. an average absorption rate of about 0.4 mg cm^{-2}h^{-1} (Gross et al., 1960). However, some evaporation could have occurred, thereby causing an overestimation of the absorption. In a study of employees at a dynamite factory, nitroglycol was detected in the blood of workers with potential skin exposure, while none was found in individuals exposed to nitroglycol vapour only (Fukuchi, 1981). Nitroglycol causes important haemodynamic changes in the body: a transient increase followed by a compensatory decrease in coronory blood flow (NIOSH, 1978). Acute effects in humans include headache, sweating, dizziness, hypotension and palpitations. These symptoms and signs tend to disappear after 3–4 days of repeated exposure, probably as a result of compensatory vasoconstriction. The exposure level below which significant changes in the diameter of blood vessels, as indicated by throbbing headaches and changes in blood pressure, is unknown. Cases of sudden death in dynamite workers occur after cessation of exposure and may be related to withdrawal effects (NIOSH, 1978). Epidemiological evidence suggests that dynamite workers suffer a considerable excess mortality from chronic cardio- and cerebrovascular disease (Hogstedt and Andersson, 1979).

Conclusion

Nitroglycol penetrates the skin rather rapidly, and skin absorption is the most important route of entry into the human body during occupational exposure. Because of the toxic effects associated with percutaneous absorption of this compound, nitroglycol should be regarded as a skin absorption hazard.

Nitroglycerin

Skin exposure potential

Nitroglycerin (1,2,3-propanetriol trinitrate) is produced by direct nitration of glycerol; information on production or consumption in EC countries is not available. Nitroglycerin is used in the manufacture of industrial explosives and as a therapeutic agent. Percutaneous application will release the drug to the circulation in a more continuous way, thus providing some benefits over ingestion (Wester, 1985).

Physicochemical properties

Nitroglycerin is an oily liquid with a melting point of 13°C; the vapour pressure at room temperature is very low. Nitroglycerin is slightly soluble in water, soluble in ethanol and miscible with acetone, diethyl ether, benzene and other organic solvents. The octanol/water partition coefficient (log P_{ow}) has been reported as 2.05 or 2.35.

Experimental data

Percutaneous absorption of nitroglycerin has been documented in rats; the absorption rate averaged 0.85 mg $h^{-1}cm^{-2}$ from a mixture with gelatin and 0.63 mg h^{-1} mc^{-2} from a mixture employed in explosives manufacture (Gross *et al.*, 1960). Thus, the absorption rate was almost independent of the nitroglycerin concentration under the experimental circumstances. The percutaneous first-pass metabolism is about 20% (Wester, 1985). Nitroglycerin is absorbed much more slowly through the skin than is ethylene glycol dinitrate, but the low vapour pressure makes a protracted absorption possible. Vasodilatation is a well documented effect in experimental studies (Rabinowitch, 1944).

Human data

In gunpowder production, excessive plasma levels of nitroglycerin were detected in blood samples from the cubital vein, as compared to much lower levels in blood from a femoral vein; these results suggested that considerable percutaneous absorption took place, despite the use of protective clothing on hands and arms (Sivertsen, 1984; Gjesdal *et al.*, 1985). Cutaneous application is utilized in certain drug preparations containing nitroglycerin; the absorption rate varies considerably, but the average is about 0.04 mg cm^{-2}h^{-1} (Curry and Aburawi, 1985), and by application of nitroglycerin-containing products on less than 16 cm^2 of skin, therapeutic concentrations can be obtained in serum (Curry and Aburawi, 1985). The factors that influence skin penetration have been studied in detail (Wester, 1985). The first sympton in exposed workers is often headache (Carmichael and Lieben, 1963; Lanfranchi and Beraud, 1969; Gjesdal et al., 1985). Individuals who handle nitroglycerin-containing materials with bare hands readily develop headache; use of a towel that a worker has used for wiping hands or face may also precipitate a headache (Rabinowitch, 1944). Excess amounts of nitroglycerin cause vasodilatation, increased heart rate and reduced blood and pulse pressure; cases of sudden death among explosives workers have probably been due to a rebound effect following vasoconstrictive adaptation to the hypotensive action (Carmichael and Lieben, 1963). Allergic contact dermatitis has also been recorded (Fisher, 1986).

Conclusions

Percutaneous absorption of nitroglycerin is well documented. Cutaneous exposure to nitroglycerin in industry may result in plasma levels much above those used in drug therapy, and adverse effects, including sudden deaths, have been linked to the effects of nitroglycerin on the blood vessels. Nitroglycerin should therefore be regarded as a skin absorption hazard.

Tetryl

Skin exposure potential

Tetryl (trinitrophenylmethylnitramine) is used as an explosive in detonators and primers. No information is available on current production or the number of workers with potential skin exposure.

Physicochemical properties

At room temperature, tetryl is a colourless or yellow solid (melting point, 130°C). It is almost insoluble in alcohol and easily soluble in ether. Thus, tetryl is lipophilic and would be expected to penetrate skin slowly, but the penetration may continue long after the initial skin contact.

Experimental data

No information is available on skin absorption of tetryl. However, experimental studies have documented that tetryl is an irritant that can cause contact dermatitis and severe airway irritation; systemic toxicity mainly affects the kidney and the liver (von Oettingen, 1941).

Human data

Occupational exposure to tetryl dust may, within several days, result in conjunctivitis and facial dermatitis. With continued exposure, the dermatitis tends to fade and many patients become less sensitive to the irritant effects (Schwartz, 1944). Contact with tetryl causes a bright yellow staining of the skin. Upper airway irritation may progress to epistaxis and severe swelling with partial obstruction of the trachea (Witkowski et al., 1942; Schwartz, 1944). Systemic effects are poorly documented, but may include anaemia, gastrointestinal symptoms and liver toxicity (NIOSH/OSHA, 1981). None of the case reports on tetryl poisoning unequivocally identifies percutaneous absorption as the cause. However, the fact that this compound penetrates the stratum corneum is documented by its severely irritant effects and the discolouration of the skin.

Conclusion

Indirect evidence indicates that tetryl may slowly penetrate the skin. The systemic toxicity potential is documented experimentally. With the strong irritant effects on the skin, tetryl should be regarded as a skin exposure hazard.

Nitrobenzene

Skin exposure potential

Within the EC, 350 000 tonnes of nitrobenzene were produced in 1980, mainly for use in organic synthesis of aniline dyes. Under the name of oil of mirbane, nitrobenzene has been used as a perfume. It is also used as a solvent and, at least in the past, as a constituent of shoe and floor polish. A potential for skin exposure exists both in the production and user industries. Thus, contamination of skin and work clothes of workers in the dye manufacturing industry may reach levels of 2 and 25 mg cm^{-2}, respectively (Trojanowska, 1959).

Physicochemical properties

Nitrobenzene is a pale-yellow liquid at room temperature and has a strong odour of bitter almonds and a vapour pressure of 0.3 mm Hg at 20°C. The boiling point is 211°C. One litre of water will dissolve 1.9 g of nitrobenzene, but the compound is very soluble in ethyl alcohol, ethyl ether and benzene. The molecular size is relatively small (molecular weight = 123). The octanol/water partition coefficient (log P_{ow}) is about 1.84. The solubility in psoriasis scales suggests a considerable dermal uptake of this compound (Hansen and Andersen, 1988). A penetration rate of 0.19 mg cm^{-2}h^{-1} was calculated on the basis of physicochemical properties (Fiserova-Bergerova and Pierce, 1989).

Experimental data

The absorption rate of nitrobenzene applied on the skin varied from 0.2 to 3 mg cm^{-2} h^{-1}, increasing with skin temperature and decreasing with duration of contact (Salmowa and Piotrowksi, 1960). Studies using human and monkey skin *in vitro*, the latter also *in vivo*, suggested that a few per cent of the applied dose was absorbed through the skin, although higher absorption was obtained *in vitro*, especially under occlusion (Bronaugh and Maibach, 1985). The lowest lethal dose applied to the skin in both the rabbit and the rat was found to be about 600 mg kg^{-1}, with an average least lethal dose of 1.5–2.0 g kg^{-1} (Czajkowska, 1982). In another study, the acute lethal dose was found to be 2.76 g kg^{-1} in rats, and 270 mg kg^{-1} for daily applications on the skin (Malysheva *et al.*, 1977). The LD$_{50}$ values for oral and dermal applications in the rat have been found to be 640 and 2100 mg kg^{-1}, respectively (Sziza and Magos, 1959). The primary toxic effect is formation of methaemoglobin in the blood, therby blocking oxygen transport in the body (ACGIH, 1986; Izmerov, 1984).

Human data

Limited information on percutaneous absorption in humans is available. Human nitrobenzene intoxications have been related to drenched work clothes in several cases (Hamilton, 1919). A study of absorbed doses of nitrobenzene showed that the total absorption considerably exceeded the possible contribution from respiratory intake (Piotrowski, 1966). Calculations suggest that exposure of a skin area of 360 cm^2 would cause an absorption almost 50-fold above the uptake from inhalation of nitrobenzene at the TLV level during the same time period (Fiserova-Bergerova and Pierce, 1989). Contact dermatitis has been produced by irritation and, in some cases, sensitization due to nitrobenzene (Fisher, 1986). The onset of nitrobenzene poisoning is insidious. In mild cases, cyanosis, headache, dizziness, hypotension, and dyspnoea may develop. In severe poisoning, hyperbilirubinaemia with anuria may ensue. Neurotoxicity is indicated by excitation and tremor followed by depression, and coma in severe poisoning. Hepatotoxicity may lead to yellow atrophy and icterus. Lethal poisoning in humans has occurred after ingestion of a few millilitres of the compound (Izmerov, 1984).

Conclusion

This compound has physical and chemical properties which suggest that skin absorption could be a hazard. Experimental results and industrial experience confirm that skin absorption is an important route of entry. Nitrobenzene should therefore be considered a skin absorption hazard.

2,4,6-Trinitrotoluene

Skin exposure potential

2,4,6-Trinitrotoluene (TNT) has been the most commonly used explosive in the munitions industries since World War I. Occupational exposure is often due to dust from the filling and packing operations.

Physicochemical properties

TNT is a yellow solid at room temperature, the melting point is 80°C. It is insoluble in water, but soluble in ethanol, ether and benzene. With a molecular weight of 227, TNT would be expected to penetrate the skin rather slowly, but absorption may continue, unless TNT is efficiently removed from the skin.

Experimental data

Little information is available in relation to skin absorption. One report stated that radiolabelled TNT was absorbed slowly after dermal application in rats (El-Hawari et al., 1978). TNT penetration of guinea pig skin in vitro showed a lag time of up to 2 h, with about 5% being absorbed within 10 h; slower rates were seen with human skin, and faster rates with mouse skin (Zi-chu et al., 1988). Toxic effects include haemolytic anaemia, peripheral neuropathy and liver toxicity (Hathaway, 1977).

Human data

Cutaneous exposure to TNT may result in contact dermatitis and a yellow discolouration of the skin (Schwartz, 1944). That penetration of the skin actually takes place to a significant extent was indicated by Woollen et al. (1986) who found urinary TNT metabolite levels much higher in explosives factory workers than could be accounted for by respiratory exposures. Reports on serious human toxicity orginate mainly from wartime conditions. All workers who developed aplastic anaemia after TNT exposure had documented or suspected skin contact with the compounds (Hathaway, 1977). A controlled study by Stewart et al. (1945) showed primarily haemolytic effects, in part, related to skin exposure to TNT. Indicators of liver toxicity have also been described (Goodwin, 1972). Much less toxicity has been reported in recent studies where respiratory exposures have been lower and dermal exposure minimized (Hathaway, 1977). Individuals with glucose-6-phosphate dehydrogenase deficiency are particularly susceptible to the haemolytic effects of TNT (Djerassi and Vitany, 1975).

Conclusions

Skin exposure to TNT has reportedly led to TNT toxicity in the past. Although the exact contribution by percutaneous absorption to the total uptake has not been documented, and although the experimental data base is very limited, prudence would suggest that TNT should be regarded as a skin absorption hazard.

Picric acid

Skin exposure potential

Picric acid (2,4,6-trinitrophenol) is used for a variety of purposes, including production of dyes and explosives, laboratory analyses and tanning purposes. Due to the explosion risk, only a very limited number of workers come into contact with picric acid, but skin contact with aqueous solutions may be more widespread.

Physicochemical properties

At room temperature, picric acid is a solid (melting point, 12°C). The water solubility is about 1.4%, and picric acid is readily soluble in ethanol, ether and benzene. The octanol/water partition coefficient is 500 (Jetzer *et al.*, 1986). Thus, with a fairly low molecular size (M = 229) and the lipophilic solubility properties, this compound would be likely to penetrate the skin rather readily.

Experimental data

Limited experimental data are available on picric acid. The *in vitro* permeability coefficient has been determined for skin from hairless mice (Huq *et al.*, 1986) and from humans (Flynn and Yalkowski, 1972) at up to 174 and 59×10^3 cm h^{-1}, respectively. Thus, picric acid penetrates the skin much more rapidly than does phenol.

Human data

Picric acid may cause both irritative and allergic contact dermatitis, especially on the face; both skin and hair may be stained yellow (Schwartz, 1944; EPA, 1980). Neurotoxic symptoms and signs are prominent in the clinical picture of intoxication, and high doses also cause haemolysis, haemorrhagic nephritis and hepatitis (Dennie *et al.*, 1929). The most serious case of systemic toxicity due to percutaneous absorption was due to the use of picric acid for topical treatment of burns; recovery generally followed rapidly after cessation of the treatment, but some intoxications have been fatal (Dennie *et al.*, 1929). Less-serious poisoning has been seen in industry following dermal exposure (EPA, 1980).

Conclusions

The limited information available strongly supports the notion that picric acid may constitute a skin absorption hazard. Although human exposures to this compound may already be controlled due to the explosion hazard, warning against skin contract with picric acid and solutions thereof would be warranted.

4,6-Dinitro-o-cresol

Skin exposure potential

Of the six isomers, only 4,6-dinitro-*o*-cresol (generally referred to as DNOC) is commercially important, although its use is declining. DNOC is primarily used as a blossom-thinning agent for fruit trees and as a pesticide applied to fruit trees during the dormant season. However, DNOC is highly and non-selectively toxic to plants in the growing stage, and it is also toxic to mammals. DNOC has limited uses for certain laboratory purposes. NIOSH (1978) has estimated that 3000 workers in the USA are potentially exposed to DNOC, mainly during spraying operations. Spraying operations may result in dermal absorption of a magnitude similar to that due to inhalation (van Noort *et al.*, 1960). Absorbent pads placed on the body of spray operators, showed an average dermal exposure of 63 mg h^{-1}; if only 1% of this amount was absorbed, the percutaneous absorption would be of the same magnitude as respiratory uptake (Batchelor *et al.*, 1956).

Physicochemical properties

DNOC is a yellow solid at room temperature; the melting point is 86°C. The solubility in water is about 0.01%, and almost 2% in ethanol (NIOSH, 1978). Due to the high lipid solubility, the very slow evaporation and the molecular size (M = 198), DNOC would be expected to penetrate skin relatively slowly, but skin absorption could be a protracted process.

Experimental data

The LD$_{50}$ in mice following dermal DNOC application is 187 mg kg^{-1}, i.e. up to 10-fold above the values obtained with other application routes (NIOSH, 1978). However, skin exposure of rats and rabbits to a 2% suspension for 30 days did not cause any discernible effects, while another study in guinea pigs showed that the LD$_{50}$ was between 200 and 500 mg kg^{-1} (NIOSH, 1978). The toxicity is due to uncoupling of oxidative phosphorylation and resulting increased metabolic rate and body temperature; mortality is higher at increased ambient temperature (NIOSH, 1978). Unfortunately, the experimental data may not be extrapolated to the human exposure situation, as humans appear to metabolize DNOC much more slowly (NIOSH, 1978).

Human data

Accidental skin application of 50 g of a solution containing 25% DNOC was fatal within 3.5 h in a 4-year-old child (NIOSH, 1978). Considerable levels of DNOC in serum were detected in two volunteers after immersion of one or both feet in a 1% suspension of DNOC for 4 h (van Noort et al., 1960). However, daily application of a 2% suspension to armpits and forearms of two volunteers for 30 days resulted in neither local nor systemic effects (NIOSH, 1978). On the other hand, occupational health experience has shown beyond doubt that percutaneous exposure is of concern. Thus, 47 workers suffered various degrees of DNOC toxicity after working in a field which had been sprayed with an aluminium salt of DNOC the previous day (NIOSH, 1978). The toxicity of DNOC results in profuse sweating, thirst, malaise, and yellow staining of skin, hair and sclera, and in serious cases, in coma and death (NIOSH, 1978). Two cases of peripheral neuropathy have been described in relation to extensive cutaneous exposure to DNOC (Stott, 1956). A WHO (1982) working group refrained from recommending an exposure limit for DNOC in air, because skin exposure occurs easily and skin absorption affects the relationship between airborne exposure and toxicity.

Conclusions

Case reports and occupational health experience convincingly document that toxic amounts of DNOC can be absorbed through the skin. Although the experimental evidence is limited and no data are available on skin penetration rates, DNOC should be regarded as a skin exposure hazard.

Other compounds

p-Nitrochlorobenzene (4-chloronitrobenzene) is a mutagen, but no published evidence has been found to suggest that skin contact is likely to cause systemic toxicity.

1,3-Dinitrobenzene is a yellow solid, sparingly soluble in water and used in the manufacture of explosives. Acute poisoning with methaemoglobinaemia due to this compound has been common in the past. Although strict evidence was not available, percutaneous uptake from dust settled on the skin was thought to be the main route of absorption (Rejsek, 1947).

More recently, a woman was exposed to a solvent that contained 0.5% 1,3-dinitrobenzene. She developed cyanosis, haemolytic anaemia, hepatomegaly and icterus. A subsequent simulation of the work situation showed that no 1,3-dinitrobenzene could be detected in the air, but the compound readily penetrated the latex goves used for protection (Ishihara et al., 1976). Although this report does not convincingly document a relation between percutaneous uptake of this compound and the adverse reactions, it indicates a need for further studies in this area.

Skin penetration of nitrotoluene is poorly documented. Likewise, evidence of percutaneous absorption of dinitrotoluene is very limited, but this compound may be a mutagen.

The herbicide dinoseb (2-*sec*-butyl-4,6-dinitrophenol) has toxic effects similar to DNOC and may be a skin hazard as well. Within 72 h after application of this compound to the skin of rats, almost complete absorption was detected at three different doses; a slightly lower degree of absorption was seen in young rats compared to adults (Shah *et al.*, 1987).

References

ACGIH (1986), *Documentation for the threshold limit values and biological exposure indices.* 5th edn. (Cincinnati: American Conference of Governmental Industrial Hygienists).
BATCHELOR, G.S., WALKER, K.C. and ELLIOT, J.W. (1956), Dinitroorthocresol exposure from apple-thinning sprays, *American Medical Association Archives of Industrial Health,* **13,** 593-6.
BRONAUGH, R.L. and MAIBACH, H.I. (1985), In vitro models for human percutaneous absorption. In *Models in Dermatology,* eds H.I. Maibach and Lowe N.J., Vol. 2, pp. 178-188 (Basel: Karger).
CARMICHAEL, P. and LIEBEN, J. (1963), Sudden death in explosives workers, *Archives of Environmental Health,* **7,** 424-39.
CURRY, S.H. and ABURAWI, S.M. (1985). Analysis, disposition and pharmacokinetics for nitroglycerin, *Biopharmaceutics and Drug Disposition,* **6,** 235-80.
CZAJKOWSKA, T. (1982), Comparative evaluation of absorption of toxic compounds through the skin of rabbits and rats, *Przeglad Lekarski,* **38,** 659-662.
DENNIE, C.C., MCBRIDE, W.L. and DAVIS, P.E. (1929), Toxic reactions produced by the application of trinitrophenol (picric acid), *Archives of Dermatology and Syphilology,* **20,** 689-704.
DJERASSI, L.S. and VITANY, L. (1975). Haemolytic episode in G6-PD deficient workers exposed to TNT, *British Journal of Industrial Medicine,* **32,** 54-58.
EL-HAWARI, A.M., HODGSON, J.R., SHIOTSUKA, R.N. and LEE, C.C. (1978), Disposition and metabolism of 2,4,6-trinitrotoluene (TNT) after oral, dermal and intratracheal dosing to rats, *Pharmacologist,* **20,** 255.
EPA (1980), *Ambient water quality criteria for nitrophenols,* (EPA 440/5-80-063), (Washington DC: U.S. Environmental Protection Agency).
FISEROVA-BERGEROVA, V. and PIERCE, J.T. (1989), Biological monitoring V. Dermal absorption, *Applied Industrial Hygiene,* **4,** F14-F21.
FISHER, A.A. (1986), *Contact dermatitis,* 3rd edn. (Philadelphia, Lea & Febiger).
FLYNN, G.L. and YALKOWSKY, S.H. (1972), Correlation and prediction of mass transport across membranes I. Influence of alkyl chain length on flux-determining properties of barrier and diffusant, *Journal of Pharmaceutical Science,* **61,** 838-52.
FUKUCHI, Y. (1981), Nitroglycol concentrations in blood and urine of workers engaged in dynamite production, *International Archives of Occupational Environmental Health,* **48,** 339-346.
GJESDAL, K., BILLE, S., BREDESEN, J.E., BJØRGE, E., HALVORSEN, B., LANGSETH, K., LUNDE, P.K.M. and SIVERTSEN, E. (1985), Exposure to glyceryl trinitrate during gun powder production, plasma glyceryl trinitrate concentration, elimination kinetics, and discomfort among production workers, *British Journal of Industrial Medicine,* **42,** 27-31.
GOODWIN, J.W. (1972), Twenty years handling TNT in a shell loading plant, *American Industrial Hygiene Association Journal,* **33,** 41.

GROSS, E., KIESE, M. and RESAG, K. (1960), Resorption von Äthylenglykoldinitrat durch die Haut, *Archives of Toxicology,* **18,** 194–199.
HAMILTON, A. (1919), Industrial poisoning by compounds of the aromatic series, *Journal of Industrial Hygiene,* **1,** 200–12.
HANSEN, C.M. and ANDERSEN, B.H. (1988), The affinities of organic solvents in biological systems, *American Industrial Hygiene Association Journal,* **49,** 301–8.
HATHAWAY, J.A. (1977), Tinitrotoluene: A review of reported dose-related effects providing documentation for a workplace standard, *Journal of Occupational Medicine,* **19,** 341–5.
HOGSTEDT, C. and ANDERSSON, K. (1979), A cohort study on mortality among dynamite workers, *Journal of Occupational Medicine,* **21,** 553–556.
HOGSTEDT, C. and STÅHL, R. (1980), Skin absorption and protective gloves in dynamite work, *American Industrial Hygiene Association Journal,* **41,** 367–372.
HUQ, A.S., HO, N.F.H., HUSARI, N., FLYNN, G.L., JETZER, W.E. and CONDIE, L. JR. (1968), Permeation of water contaminative phenols through hairless mouse skin, *Archives of Environmental Contamination and Toxicology,* **15,** 557–66.
ISHIHARA, N., KANAYA, A. and IKEDA, M. (1976), m-Dinitrobenzene intoxication due to skin absorption, *International Archives of Occupational and Environmental Health,* **36,** 161–8.
IZMEROV, N.F., ed. (1984), *Nitrobenzene* (Scientific Reviews of Soviet Literature in Toxicity and Hazards of Chemicals, 51) (Moscow: Centre of International Projects).
JETZER, W.E., HUQ, A.S., HO, N.F.H., FLYNN, G.L., DURAISWAMY, N. and CONDIE, L. (1986), Permeation of mouse skin and silicone rubber membranes by phenols: Relationship to in vitro partitioning, *Journal of Pharmaceutical Science,* **75,** 1098–1103.
LANFRANCHI, A. and BERAUD, P. (1969), De l'intoxication chronique par les produits nitrés chez les travailleurs des usines d'explosifs, *European Journal of Toxicology,* **2,** 191–6.
MALYSHEVA, M. V., SERGEYEV, A.N. and YUSHKEVICH, I.B. (1977), Toxicity of some dissolvents under the administration through the skin. In *Skin pathways of industrial poisons into the organism and their preventive treatment* (in Russian) (Transactions of the Erisman Hygiene Research Institute), pp. 46–50.
NIOSH (1978), *Criteria for a recommended standard... Occupational exposure to dinitro-ortho-cresol,* (DHEW (NIOSH) publication No. 78-131) (Cincinnati, OH: National Institute for Occupational Safety and Health).
NIOSH (1978), *Criteria for a recommended standard... Occupational exposure to nitroglycerin and ethylene glycol dinitrate,* (DHEW publication No. (NIOSH) 78-167) (Cincinnati, Ohio: National Institute for Occupational Safety and Health).
NIOSH/OSHA (1981), *Occupational health guidelines for chemical hazards,* (Cincinnati, Ohio: National Institute for Occupational Safety and Health).
VAN NOORT, H.R., MANDEMA, E., CHRISTENSEN, E.K.J. and HUIZINGA, T. (1960), Dinitro-ortho-cresol poisoning due to agricultural sprays (in Dutch), *Nederlands Tijdschrift voor Geneeskunde,* **104,** 676–84.
VON OETTINGEN, W.F. (1941), *The aromatic amino and nitro compounds, their toxicity and potential dangers, a review of the literature.* (Public Health Service Bulletin No. 271) (Washington, D.C.: U.S. Government Printing Office).
PIOTROWSKI, J. (1966), Chemical problems of the industrial toxicology of nitrobenzene, *Medycyna Pracy,* **17,** 519–534.
RABINOWITCH, I.M. (1944), Acute nitroglycerin poisoning, *Canadian Medical Association Journal,* **50,** 199–202.
REJSEK, M. (1947) m-Dinitrobenzene poisoning, mobilization by alcohol and sunlight, *Acta Medica Scandinavica,* **127,** 179–91.

SALMOWA, J. and PIOTROWSKI, J. (1960), Attempt on the quantitative estimation of the nitrobenzene resorption in experimental conditions (in Polish), *Medycyna Pracy,* **11**, 1-14.
SCHWARTZ, L. (1944), Dermatitis from explosives, *Journal of the American Medical Association,* **12**, 186-90.
SHAH, P.V., FISHER, H.L., SUMLER, M.R., MONROE, R.J. CHERNOFF, N. and HALL, L.L. (1987) Comparison of the penetration of 14 pesticides through the skin of young and adult rats, *Journal of Toxicology and Environmental Health,* **21**, 353-66.
SIVERTSEN, E. (1984), Glyceryltrinitrate as a problem in industry, *Scandinavian Journal of Clinical & Laboratory Investigation,* **173**, (Suppl.) 81-4.
STEWART, A., WITTS, L.T., HIGGINS, G. and O'BRIEN, J.R.P. (1945), Some early effects of exposure to tinitrotoluene, *British Journal of Industrial Medicine,* **2**, 74-82.
STOTT, H. (1956), Polyneuritis after exposure to dinitro-ortho-cresol, *British Medical Journal,* **1**, 900-1.
SZIZA, M. and MAGOS, L. (1959), Toxicological investigations of some aromatic nitro compounds used in the Hungarian industry, *Archiv für Gewerbepathologie und Gewerbehygiene,* **17**, 217.
TROJANOWSKA, B. (1959), The pollution with nitro and amino compounds of the work clothes and the skin of the dye industry workers (in Polish), *Medycyna Pracy,* **10**, 387-392.
WESTER, R.C. (1985), In vivo skin and nitroglycerin transdermal delivery. In *Percutaneous absorption,* eds R.L. Bronaugh, and H.I. Maibach, pp. 541-6. (New York: Marcel Dekker)
WHO (1982), *Recommended health-based limits in occupational exposure to pesticides,* (Technical Report Series No. 677). (Geneva: World Health Organization).
WITKOWSKI, L.J., FISCHER, C.N. and MURDOCK, H.D. (1942), Industrial illness due to tetryl, *Journal of the American Medical Association,* **119**, 1406-9.
WOOLLEN, B.H., HALL, M.G., GRAIG, R. and STEEL, G.T. (1986), Trinitrotoluene: assessment of occupational absorption during manufacture of explosives, *British Journal of Industrial Medicine,* **43**, 465-73.
ZI-CHU, G., QUI-PING, G. ZHEN-HUA, Z. BO-YUAN, Y. and TI-LAN, W. (1988), Studies on the percutaneous absorption of trinitrotoluene. In *Occupational health in industrialization and modernization,* (Bulletin of WHO Collaborating Center for Occupational Health, Vol.2), eds. X. Shou-zheng, and L. You-xin, pp. 65-69 (Shanghai: Shanghai Medical University Press).

Chapter 10
Other nitrogen compounds

Introduction

The nitrogen compounds (not considered in other chapters) most frequently considered to be skin absorption hazards are indicated in table 10.1. The most important of these are considered separately below. These compounds represent a range of different effects and various degrees of skin absorption hazards. Some are solvents, some pesticides and some are mainly intermediates for organic synthesis.

Among the heterocyclic solvents belonging to this group, morpholine, pyrrolidine, pyridine, and quinoline have solubility characteristics approaching those of psoriasis scales, and a considerable uptake in the skin would therefore be

Table 10.1. Other nitrogen compounds that are considered a skin hazard in several countries (y = yes, n = no, o = other regulation). Those that are described in separate sections are italicized.

CAS No.	Chemical	Number of countries	FRG	Sweden	USA
Other aliphatic and isocyclic nitrogen compounds					
624-83-9	Methyl isocyanate	6	o	n	y
4098-71-9	Isophorone diisocyanate	6	o	o	y
16752-77-5	Methomyl	6	n	n	y
63-25-2	*Carbaryl*	3	y	n	o
Heterocyclic nitrogen compounds					
151-56-4	Ethyleneimine	10	y	o	y
75-55-8	Propyleneimine	6	y	o	y
54-11-5	*Nicotine*	11	y	n	y
1910-42-5	*Paraquat*	7	y	n	n
2425-06-1	Captafol	6	n	n	y
110-91-8	Morpholine	10	y	y	y
100-74-3	*N*-Ethylmorpholine	7	o	y	y
Hydrazines					
302-01-2	*Hydrazine*	11	y	y	y
60-34-4	Methylhydrazine	8	o	n	y
57-14-7	*1,1-Dimethylhydrazine*	9	y	y	y
100-63-0	*Phenylhydrazine*	9	y	n	y

expected (Hansen and Andersen, 1988). N-Methylmorpholine has a 'skin' denotation in one country.

In addition to those compounds shown in the table, only two carbamates have a 'skin' denotation, each only in a single country: ferbam and carbofuran. Amitraz, a triazine herbicide, is also considered to be a skin absorption hazard in one country.

An isocyanate other than those mentioned in the table, methylenebis(4-cyclohexyl isocyanate) (bis(4-isocyanatocyclohexyl)methane), is considered to be a skin hazard in one of the countries surveyed. Similar compounds are well known causes of allergic reactions.

An additional hydrazine compound, 1,2-dimethylhydrazine, is considered a skin hazard in three countries; it is a possible carcinogen (class 2B) (IARC, 1987).

Carbaryl

Skin exposure potential

Carbaryl (1-naphthyl-N-methylcarbamate) is in common use within the EC countries, although the exact quantitites are unknown, and it is mainly used on crops, such as cotton, maize, fruits and vegetables. Occupational exposures to carbaryl occur during manufacturing and formulation processes and as a result of crop spraying and dusting. Dermal exposure is of major importance. Thus, in a study of formulation plant workers, the dermal exposure to carbaryl was estimated at 74 mg h^{-1}, while only 0.09 mg was inhaled per hour (Cromer et al., 1975). Similar results have been reached in pesticide applicators, with most dermal exposure being to forearms and hands; only about 5% of carbaryl was found to penetrate the work clothes (Leavitt et al., 1982). In a study of tractor drivers, pesticide spraying resulted in higher respiratory exposures and lower skin exposures, but the inhalation still constituted a relatively small exposure (Jegier, 1974). The exposure conditions for pesticide applicators vary considerably with, e.g. work practices, wind speed and humidity (Leavitt et al., 1982; Jegier, 1974; Zweig et al., 1985).

Physicochemical properties

Carbaryl is an odourless, white, solid material at room temperature. The melting point is 142°C, and the vapour pressure at 40°C is 0.002 mm Hg. This pesticide is formulated as wettable powders, pellets, granules and various mixtures and solution. Carbaryl is barely water soluble, and 1 l of water will only dissolve 40 mg at 30°C. It dissolves more readily in polar organic solvents, such as acetone and dimethylformamide, but is less soluble in non-polar solvents. Thus, the partition coefficient in chloroform/water has been determined to be more than 200, i.e. about six fold greater than the partition coefficient in olive oil/water (Shah et al., 1981). The octanol/water partition coefficient (log P_{ow}) is 2.34. Although the

Experimental data

Acute toxicity in rats depends on the administration route. Oral LD_{50} values have varied between 500 and 850 mg kg^{-1} (Carpenter et al., 1961; Gaines, 1969). Skin application of 500 mg kg^{-1} in rabbits resulted in enzyme inhibition of 35-40% (Yakim, 1967), an effect of limited long-term pathological significance, and the dermal LD_{50} in rats has been found to be above 4000 mg kg^{-1} (Gaines, 1969). When radioactively labelled carbaryl dissolved in acetone was applied to the shaved skin of mice, half of the carbaryl appeared to penetrate the skin in less than 15 min (Shah et al., 1981). Studies in rats (O'Brien and Danelley, 1965) and rabbits (Shah and Guthrie, 1977) have also shown that carbaryl is one of the pesticides that penetrates the fastest through the skin. The aqueous wettable powder formulation was reported to penetrate faster than did an acetone solution of carbaryl (Shah et al., 1981). The half-time for skin penetration in rats was estimated to be less than 5 h by the direct method, where the treated skin was excised and analyzed, and slightly more than 10 h by the indirect method based on urine analysis (Shah and Guthrie, 1983). Within 72 h after application of an acetone solution of carbaryl to the skin of adult rats, about 30% of the low dose, but only 3% of a 17-fold higher dose had penetrated the skin (Shah et al., 1987). Carbaryl appears to be a weak skin sensitizer (Carpenter et al., 1961). After skin penetration, this insecticide exhibits the toxic effects characteristic of carbamates and mediated by inhibition of cholinesterase. The enzyme inhibition leads to accumulation of acetylcholine that results in symptoms, such as headache, nausea, vomiting, abdominal cramps and dimness of vision. The following signs may be seen in experimental studies: salivation, lacrimation, urination, defaecation, muscle tremors, incoordination and convulsions.

Human data

Radioactively labelled carbaryl was applied to the ventral forearm of human volunteers, and the radioactivity of urine during the following 5 days was compared to the results obtained after intravenous injection of the substance. This method may lead to somewhat inaccurate results in humans, because only 8% of injected radioactivity was recovered in the urine with a half-life of about 9 h; the estimated absorption averaged about 79% for the six subjects examined, thus suggesting an almost complete absorption (Feldman and Maibach, 1974). After injection, urinary excretion rapidly reaches a maximum, but the urinary radioactivity after skin application was delayed by about 8 h. Although the penetration rate may depend on the doses applied and the solvent, this observation would suggest that human skin may be less permeable to carbaryl than is the shaved skin of mice and rats. Human toxicity is infrequently seen, but most cases appear to be caused by skin absorption (WHO, 1982). The dose–response relationship in humans is not known in any detail. The cholinesterase inhibition is reversible, and the toxicity in pesticide sprayers and formulators normally wears off before the next working day. Long-term effects have not been demonstrated in humans, although some experimental animal studies suggest that delayed effects in the reproductive system may occur (WHO, 1982).

Conclusion

Carbaryl may penetrate the skin in significant quantities over several hours. Toxic effects of cutaneous exposures have been demonstrated in rodents, and skin exposure appears to be the main cause of human cases of carbaryl intoxication.

Nicotine

Skin exposure potential

Nicotine is produced by steam distillation or by extraction with trichlorethylene from waste tobacco. In the USA, the annual consumption of nicotine as an insecticide is 500 tonnes (ECDIN, pers. comm., 1987). Nicotine has found wide use as an insecticide (particulary when pests have developed resistance against other compounds); dilute solutions are used on fruit and vegetables. Also, it may be used to control skin parasites in animals.

Physicochemical properties

Nicotine is a liquid at room temperature (boiling point, 246°C); it is miscible with water and most organic solvents. The vapour pressure at room temperature is 0.04 mm Hg. Nicotine has an octanol/water partition coefficient (log P_{ow}) of 1.17, thus suggesting a combination of hydrophilic and lipophilic properties. Although the molecular size (M = 162) would tend to limit percutaneous absorption, the physicochemical properties would suggest a relatively rapid penetration.

Experimental data

A relatively fast rate of percutaneous absorption has been documented. When applied in an acetone solution to the shaved skin of mice, half of the ^{14}C-nicotine was found to have penetrated the skin by 18 min, and the concentration in blood at 15 min was higher than for several other pesticides studied (Shah *et al.*, 1981). In a study using adult rats (Shah *et al.*, 1987), a total of 0.014 μmol cm^2 (2.3 μg cm^{-2}) was absorbed through the skin in 72 h after a single application of nicotine in an acetone solution; 75–86% of the nicotine applied was absorbed during this period. Although the penetration rate at steady state cannot be estimated, these data document that skin exposure may be followed by a virtually complete absorption. Nicotine acts on the central nervous system, first as a stimulant, then as a depressant.

Human data

Quantitative data on the rate of percutaneous absorption in humans are not available. However, case reports confirm that rapid penetration occurs. Thus, a florist sat down on a bench on which a 40% solution of nicotine had been spilled, and an area about the size of a palm was drenched; fifteen minutes later the first symptoms of nicotine poisioning occured, but full recovery took several days (Faulkner, 1933). Prolonged absorption despite vigorous skin decontamination after cutaneous exposure to a 40% solution of nicotine sulfate was documented in a patient who had used the solution against scabies; tolerance developed, and the symtoms disappeared despite continuing high plasma concentrations of nicotine (Benowitz et al., 1987). Cases of poisoning have occurred in the past when using nicotine as a pesticide; brown stains on the skin may indicate the aetiology (Wilson, 1930). A high standard of vigilance is recommended to prevent injurious contact with this highly toxic substance (ILO, 1983). Hazardous amounts of nicotine may be absorbed through the skin when harvesting green tobacco; the intoxication is self-limited and symptoms include nausea, vomiting, dizziness and prostration (Gehlbach et al., 1975). Waterproof protective clothing has prevented nicotine absorption in tobacco pickers (Gehlbach et al., 1979).

Conclusions

Experimental data document that nicotine can be absorbed through the skin. Occupational experiences relate to percutaneous absorption of nicotine used as a pesticide or from wet tobacco leaves. The rapid, but prolonged, uptake and the systemic effects encountered document that nicotine is a skin exposure hazard.

Paraquat

Skin exposure potential

Paraquat is produced from pyridine and is used as a pesticide, herbicide and defoliant. Cutaneous exposure may occur during production, formulation, spraying and as a result of skin contact with sprayed produce. In paraquat application, skin exposures from hand-held equipment have generally occurred at a rate of up to several milligrammes per hour (Hart, 1987; WHO, 1984). Significant skin exposures can be prevented during spraying operations, but systemic absorption may particularly occur in association with minor skin abrasions on hands or legs where absorption is faster (Swan, 1969).

Physicochemical properties

Paraquat is generally used as a salt, such as the dichloride: the pure compound is a colourless, hygroscopic powder. Paraquat has a melting point of 175-180°C and is involatile at ambient temperature. Paraquat is very soluble in water, slightly soluble in alcohols and insoluble in hydrocarbons.

Experimental data

The dermal LD_{50} for male rats is $80\,mg\,kg^{-1}$, for female rats $90\,mg\,kg^{-1}$, slightly below the oral LD_{50} values (Gaines, 1969). Thus, dermal exposure is as toxic as oral exposure. The dermal LD_{50} in rabbits has been reported as $236\,mg\,kg^{-1}$ (single dose), or $4.5\,mg\,kg^{-1}\,day^{-1}$ for 20 days (Clark et al., 1966). For human skin in vitro, the permeability constant for an aqueous solution of paraquat was $0.73 \times 10^5\,cm\,h^{-1}$, much below the permeability seen with other species (Walker et al., 1983). Paraquat is a skin irritant, but not a sensitizer (WHO, 1984). Signs of acute poisoning suggest neurotoxic effects, but consolidation and oedema of the lungs were the apparent cause of death (Clark et al., 1966; Kimbrough and Gaines, 1970). Low dermal absorption over a 9-week period resulted in pulmonary artery lesions in the rat (Levin et al., 1979). The toxic mechanism appears to be induction of peroxidation in the lungs (WHO, 1984).

Human data

Using ^{14}C-labelled paraquat without occlusion, percutaneous absorption was demonstrated in human volunteers at a low rate of $0.03\,\mu g\,cm^{-2}$ for a 24-h exposure, as compared to a rate of $0.5\,\mu g\,cm^{-2}$ in vitro; a higher rate of absorption would be likely to occur under occlusion and in case of damaged skin (Wester et al., 1984). Despite the extensive use of this pesticide, human intoxication appears to be relatively rare (Pasi, 1978). Systemic toxicity, sometimes fatal, due to paraquat particularly occurs due to relatively concentrated solutions (Tungsanga et al., 1983; Jaros, 1978; Okonek et al., 1983). Fatal cases have been recorded as a result of extensive skin exposure during spraying operations due to leaking equipment (Wohlfahrt, 1982) or failure to use protective equipment (Newhouse et al., 1978). In contrast, no fatal case has been reported as a result of inhalation exposure (WHO, 1984). Paraquat has a strong irritant effect on the skin and may cause nail damage; systemic toxicity may lead to pulmonary fibrosis which develops over a period of several days after a severe, acute exposure, and renal failure and multiple organ toxicity may also occur (Vale et al., 1987; WHO, 1984).

Conclusions

Percutaneous absorption of paraquat may cause systemic toxicity in experimental animals. However, paraquat penetrates human skin very slowly. Human cases of paraquat poisoning have been related to skin contact with solutions of relatively high concentrations of paraquat. Paraquat should therefore be regarded as a skin absorption hazard.

Hydrazine

Skin exposure potential

Hydrazine may be produced by several methods. Its main uses are as an anticorrosive water additive for cooling systems and boiler feeding and for rocket fuel, as a polymerization catalyst, for the synthesis of blowing agents, and as an intermediate agent for the production of several pesticides and drugs. The annual production and consumption within the EC countries is 8–12 thousand tonnes. Hydrazine is used as a rocket propellant, polymerization catalyst, oxygen scavenger in boiler water treatment, and as an intermediate in organic synthesis. In the USA, approximately 9000 workers are potentially exposed to hydrazine and more than 90 000 workers are exposed to hydrazine salts (NIOSH, 1978).

Physicochemical properties

Hydrazine is an oily liquid with a boiling point of 113.5°C and a vapour pressure of 14.4 mm Hg at 25°C. It is miscible with water and alcohol, but insoluble in chloroform, ether and hydrocarbons; it dissolves many inorganic substances. Hydrazine has a negative log P_{ow} of about -1, thus confirming the high degree of hydrophilicity. The small molecular size of this non-polar compound would suggest a potential for rapid skin penetration.

Experimental data

The dermal LD_{50} in rabbits has been reported as 91 mg kg^{-1} (Krop, 1954) and 93 mg kg^{-1} (WHO, 1987), i.e. about threefold above the value obtained by oral or intravenous administration. When hydrazine was applied directly to shaved chest skin of mongrel dogs, systemic absorption could be detected in blood samples from the femoral artery within 30 s, and maximum concentrations were seen after 1–3 h (Smith and Clark, 1972). Similar results were obtained in a study of rabbits where an aqueous solution of hydrazine (700/l) was applied to an area of shaved skin without occlusion; taking into account the evaporation loss, 55% of the dose was absorbed through the skin, and an absorption constant (K_a) of 3.47 h^{-1} was calculated (Andersen and Keller, 1984). Upon cutaneous application, the skin turned red with 2 min and then progressed to oedema which subsided within 30 min (Smith and Clark, 1972). A cutaneous dose of two droplets was rapidly fatal to rats, and 3 ml to rabbits (NIOSH, 1978). Mortality was recorded in dogs as dermal doses of 3 m mol kg^{-1} (96 mg kg^{-1}) and above (Smith and Clark, 1972). Convulsions, liver toxicity and haemolytic anaemia are the main effects observed in experimental animals (NIOSH, 1978). Sufficient evidence is available for carcinogenicity to animals (WHO, 1987; IARC, 1987).

Human data

At an industrial hydrazine explosion, a worker suffered thermal burns, CNS toxicity developed, and after 14 h he became comatose (Kirklin et al., 1976). Percutaneous uptake was probably increased due to the extensive burns. In a fatal case of hydrazine poisoning, air levels were low, and uptake through the skin was probably the main route of absorption (Sotaniemi et al, 1971). Skin irritation from hydrazine and its various salts is pronounced, and direct contact may result in burns or sensitization (Evans, 1959; Frost and Hjorth, 1959; Høvding, 1967; Keilig and Speer, 1983; Wheeler et al., 1965). The CNS is the target organ in acute exposure (WHO, 1987). In predisposed individuals with slow acetylation, occupational hydrazine exposure may lead to the development of a disease resembling systemic lupus erythematosus (Reidenberg et al., 1983). Further, while human data are inadequate, hydrazine is considered a possible human carcinogen, i.e. class 2B (IARC, 1987).

Conclusions

Hydrazine very rapidly penetrates the skin, and skin contact may cause considerable uptake. Prolonged contact may also result in sensitization. The main adverse effects are due to neurotoxicity in acute exposures and possible carcinogenicity. Hydrazine should therefore be regarded as a skin exposure hazard.

1,1-Dimethylhydrazine

Skin exposure potential

1,1-Dimethylhydrazine (unsymmetrical dimethylhydrazine, UDMH) may be produced by reduction of *N*-nitrosodimethylamine or by reaction of dimethylamine with chloramine. No information is available on production in Europe. Dimethylhydrazine is used as a rocket propellant and jet fuel and as an intermediate for production of, e.g. drugs and pesticides (NIOSH, 1978). In the USA, 1500 workers are potentially exposed to this compound (NIOSH, 1978).

Physicochemical properties

1,1-Dimethylhydrazine is a hygroscopic liquid with a boiling point of 63°C and a vapour pressure of 157 mmHg at 25°C. It is miscible with water, ethanol, ethyl ether, dimethylformamide and hydrocarbons (NIOSH, 1978). The octanol/water partition coefficient (log P_{ow} is approx. -1.9) suggests that this compound is primarily hydrophilic, and as a non-polar compound of relatively small molecular size it would be expected to penetrate the skin quite readily.

Experimental data

The dermal LD_{50} in dogs (1.2–1.7 g kg^{-1}) is about 10-fold above that seen with other administration routes (Smith and Clark, 1971). Similar dermal LD_{50} values were observed in rabbits and guinea pigs (Smith and Clark, 1971). When 1,1-dimethylhydrazine was applied directly to shaved chest skin of mongrel dogs, systemic absorption could be detected in blood samples from the femoral artery within 30 s, but no further increase was seen 5 min later, and maximum concentrations occurred after 1 h or more (Smith and Clark, 1971). A slight reddening within 10–15 minutes was the only indication of skin irritation; all dogs given 30 mmol (1.8 g) of the compound cutaneously died, but mortality was also seen at 5 mmol (300 mg) (Smith and Clark, 1971). The main adverse effects are convulsions, salivation, vomiting and haemolytic anaemia; 1,1-dimethylhydrazine may also by hepatotoxic (NIOSH, 1978). Sufficient evidence for carcinogenicity has been documented in experimental animal studies (IARC, 1987).

Human data

Skin contact may result in corrosive effects. Limited case reports suggest that 1,1-dimethylhydrazine is a central nervous system poison and may cause tremor, clonic movements, convulsions and dyspnoea (Frierson, 1965). Also, liver toxicity and haemolytic anaemia may occur (Shook and Cowart, 1957). Increased liver enzymes were seen in personnel working with rocket propellants, although the relation to 1,1-dimethylhydrazine exposure was not proven beyond doubt (Petersen et al., 1970). The contribution by skin-absorption to systemic toxicity has not been studied in detail. Insufficient epidemiological data are available to assess the carcinogenic potential for 1,1-dimethylhydrazine in humans, and it is considered a class 2B carcinogen (IARC, 1987).

Conclusions

Experimental studies show that 1,1-dimethylhydrazine rapidly penetrates the skin and may result in systemic toxicity. This compound is a possible carcinogen. Detailed information on the significance of skin absorption in humans is unavailable. However, prudence would suggest that 1,1-dimethylhydrazine should be regarded as a skin absorption hazard.

Phenylhydrazine

Skin exposure potential

Phenylhydrazine is primarily produced by reduction of diazobenzene with stannous chloride. No production or consumption figures for Europe are available. Phenylhydrazine is used as an analytical reagent, a catalyst and a blowing agent in rubber processing; it is also used as an intermediate in the production of various herbicides, dyes and pharmaceuticals. Phenylhydrazine hydrochloride also has limited use as a therapeutic agent. In the USA, about 5000 workers are potentially exposed to this compound (NIOSH, 1978).

Physicochemical properties

With a melting point of 20°C, phenylhydrazine may appear as a crystalline substance or an oily liquid. The vapour pressure is low, only 1 mm Hg at 72°C. Phenylhydrazine is sparingly soluble in water, but soluble in dilute acids and in ethanol, ethyl ether, acetone and benzene. The octanol/water partition coefficient (log P_{ow}) is 1.25, thus indicating some lipophilicity. The low vapour pressure and the solubility characteristics would suggest that phenylhydrazine penetrates the skin quite readily.

Experimental data

No quantitative data on skin absorption have been identified, but phenylhydrazine is known to be readily absorbed through the skin (Anon., 1942). Phenylhydrazine is strongly haemolytic and causes severe anaemia, bilirubinaemia, urobilinuria, icterus and liver damage (Anon., 1942; NIOSH, 1978). Carcinogenicity of phenylhydrazine has been documented in mice (NIOSH, 1978).

Human data

Toxic effects related to occupational exposures are mostly due to direct skin contact; the adverse effects recorded include fatigue, headache, vertigo, digestive disturbances, anaemia and dyspnoea (Anon., 1942; Schuckmann, 1969). From therepeutic use of phenylhydrazine, several adverse effects have been recorded, including haemolytic anaemia, liver damage and occasional kidney dysfunction and urticaria (Giffin and Conner, 1929; Solomons, 1946). Skin contact with phenylhydrazine or its derivatives may result in severe burns; sensitization occurs in a high proportion of those exposed, and cross-sensitization is frequent (Schuckmann, 1969; Pevny and Peter, 1983; Rothe, 1988). No information is available on carcinogenicity in humans.

Conclusions

Phenylhydrazine has physicochemical properties that suggest that it may penetrate the skin quite rapidly. No quantitative data are available, but industrial experience indicates that adverse effects are mainly caused by skin contact. With the documented systemic toxicity and the carcinogenic potential in experimental animals, phenylhydrazine should be regarded as a skin absorption hazard.

Other compounds

Methyl isocyanate has a high vapour pressure (348 mm Hg at 20°C), but skin contact can cause irritation and sensitization. An approximate dermal LD_{50} for methyl isocyanate in rabbits was found to be 0.22 ml kg^{-1} (213 mg kg^{-1}) (Smyth et al., 1969). A more recent study suggested a dermal LD_{50} in rabbits of about 1800 mg kg^{-1} (Vernot, 1977). Toxic effects are primarily due to inhalation of the vapours (McConnell et al., 1987). However, based on the data given by Smyth et al., 1969). ACGIH (1986) considered a 'skin' denotation 'warranted, as sufficient methyl isocyanate can penetrate rabbit skin to cause death'.

Toluene-2,4-diisocyanate is also a sensitizer and skin irritant. Urinary excretion of the metabolite toluenediamine has been used to quantify total absorption, including percutaneous uptake (Rosenberg and Savolainen, 1986). Toluene diisocyanates are classified as possible carcinogens (IARC, 1987).

Isophorone diisocyanate (3-isocyanatomethyl-3,5,5-trimethylcyclohexylisocyanate) has also caused cases of allergic contact dermatitis (Rothe, 1976).

Methomyl (methyl-N-((methylcarbamoyl)oxy)thioacetimidate) is a pesticide with possible mutagenic effects. When applied in an acetone solution to the shaved skin of mice, half of the ^{14}C-methomyl was found to have penetrated the skin within several minutes, and the concentration in blood at 15 min was higher than for several other pesticides studied (Shah et al., 1981).

A fatal case of ethyleneimine poisoning was attributed to accidental dermal exposure of the stomach and legs with 150 ml of the compound; some inhalation exposure may have occurred. A severe chemical burn developed, as well as systemic toxicity, and the patient died 19 days after the exposure (Thiess, 1965). The dermal LD_{50} for this compound in the guinea pig is 14 mg kg^{-1}; for propyleneimine, the value is 43 mg kg^{-1} (Smyth and Carpenter, 1948). The imines are primary skin irritants; experimental data suggest that both ethyleneimine and propyleneimine may be carcinogenic.

Captafol is a well established cause of allergic contact dermatitis (Brown, 1984), and it is suspected of being a carcinogen.

Within 72 h after application of carbofuran, a carbamate insecticide, to the skin of adult rats, an almost complete absorption was detected at the lowest dose, but less than 10% of a 100-fold higher dose was absorbed; young animals showed a lower degree of absorption than did adults (Shah et al., 1987). Aldicarb has a very low dermal LD_{50}, though somewhat above the oral value (Gaines, 1969). The carbamates reversibly inhibit cholinesterase. Toxicity has been documented in humans, but little information is available on the significance of skin absorption of carbamates other than carbaryl.

For captan, a common phthalimide fungicide, percutaneous absorption in rats after 72 h was about 4% at high doses, though about 10-fold higher at a very small dose (Shah et al., 1987). This pesticide can cause allergic contact dermatitis (Fisher, 1986) and is suspected of being carcinogenic. Another phthalimide fungicide, folpet, showed a lower absorption of 1–3% at the highest doses after 72 h in the same study (Shah et al., 1987).

The same authors (Shah et al., 1987) also studied percutaneous absorption of atrazine, a widely used triazine herbicide, in rats; a maximal absorption of about 10% was seen at 72 h with the lowest dose, but a decreased relative absorption was evident with higher doses.

Morpholine and *N*-ethylmorpholine are primary irritants that are also suspected of possible carcinogenicity. The dermal LD_{50} in rabbits is $0.5\,\mathrm{ml\,kg^{-1}}$ for morpholine (Smyth et al., 1954).

References

ACGIH (1986), *Documentation for the threshold limit values and biological exposure indices*, 5th edn. (Cincinnati: American Conference of Governmental Industrial Hygienists).

ANDERSEN, M. E. and KELLER, W. C. (1984), Toxicokinetic principles in relation to percutaneous absorption and cutaneous toxicity. In *Cutaneous toxicity*, eds. V.A. Drill and P. Lazar, (New York: Raven Press). pp. 9–27.

ANON (1942), Phenylhydrazine and its derivatives, *Public Health Bulletin*, **271**, 158–71.

BENOWITZ, N. L., LAKE, T., KELLER, K. H. and LEE, B L. (1987), Prolonged absorption with development of tolerance to toxic effects after cutaneous exposure to nicotine, *Clinical Pharmacology and Therapeutics*, **42**, 119–20.

BROWN, R. (1984), Contact sensitivity to Difolatan (Captafol), *Contact Dermatitis*, **10**, 181–2.

CARPENTER, C. P., WEIL, C. S., PALM, P. E., WOODSIDE, M. W., NAIR, J. H., III and SMYTH, H. F., JR (1961), Mammalian toxicity of 1-naphthyl-*N*-methylcarbamate (Sevin insecticide), *Journal of Agricultural Food Chemistry*, **9**, 30–39.

CLARK, D. G., McELLIGOTT, T. F., and HURST, E. W. (1966). The toxicity of paraquat *British Journal of Industrial Medicine*, **23**, 126–32.

COMER, S. W., STAIFF, D. C., ARMSTRONG, J. F. and WOLFE, H. R. (1975), Exposure of workers to carbaryl, *Bulletin of Environmental Contamination and Toxicology*, **13**, 385–391.

ECDIN (1987). Pers. comm.

EVANS, D. M. (1959), Two cases of hydrazine hydrate dermatitis without systemic intoxication, *British Journal of Industrial Medicine*, **16**, 126–7.

FAULKNER, J. M. (1933), Nicotine poisoning by absorption through the skin, *Journal of the American Medical Association*, **100**, 1664–5.

FELDMAN, R. J. and MAIBACH, M. I. (1974), Percutaneous penetration of some pesticides and herbicides in man, *Toxicology and Applied Pharmacology*, **28**, 126–132.

FISHER, A. A. (1986), *Contact dermatitis*, 3rd edn. (Philadelphia, Lea & Febiger).

FRIERSON, W. B. (1965), Use of pyridoxine HCl in acute hydrazine and UMDH intoxication, *Industrial Medicine and Surgery*, **34**, 650–1.

FROST, J. and HJORTH, N. (1959), Contact dermatitis from hydrazine hydrochloride in soldering flux, cross sensitization to apresoline and isoniazid, *Acta Dermato-Venereologica*, **39**, 82–6.

GAINES, T. B. (1969), Acute toxicity of pesticides, *Toxicology and Applied Pharmacology*, **14**, 515–534.

GEHLBACH, S. H., PERRY, L. D., WILLIAMS, W. A., FREEMAN, J. I., LANGONE, J. J. PETA, L. V., and VAN VUNAKIS, H. (1975). Nicotine absorption by workers harvesting green tobacco, *Lancet*, **i**, 478–80.

GEHLBACH, S. H., WILLIAMS W. A. and FREEMAN, J. I. (1979), Protective clothing as a means of reducing nicotine absorption in tobacco harvesters, *Archives of Environmental Health*, **34**, 111-4.

GIFFIN, H. Z. and CONNER, H. M. (1929), The untoward effects of treatment by phenylhydrazine hydrochloride, *Journal of the American Medical Association*, **92**, 1505-7.

HANSEN, C. M. and ANDERSEN, D. H. (1988), The affinities of organic solvents in biological systems, *American Industrial Hygiene Association Journal*, **49**, 301-8.

HART, T. B. (1987), Paraquat — a review of safety in agricultural and horticultural use, *Human Toxicology*, **6**, 13-8.

HØVDING, G. (1967), Occupational dermatitis from hydrazine hydrate used in boiler protection, *Acta Dermato-Venereologica*, **47**, 293-7.

IARC (1987), *Overall evaluations of carcinogenicity: An updating of IARC Monographs, Volumes 1 to 42*, (IARC Monographs on the evaluation of carcinogenic risk of chemicals to man, (Suppl. 7.)) (Lyon: International Agency for Research on Cancer).

ILO (1983), *Encyclopaedia of occupational health and safety*, 3rd edn. (Geneva: International Labour Office).

JAROS, F. (1978), Acute percutaneous paraquat poisoning, *Lancet*, **i**, 275.

JEGIER, Z. (1974), Health hazards in insecticide spraying of crops, *Archives of Environmental Health*, **8**, 670-674.

KEILIG, W. and SPEER, U. (1983), Berufsbedingte allergische Kontaktdermatitis aus Hydrazin, *Dermatosen in Beruf und Umwelt*, **31**, 25-7.

KIMBROUGH, R. D. and GAINES, T. B. (1970), Toxicity of paraquat to rats and its effect on rat lungs, *Toxicology and Applied Pharmacology*, **17**, 679-90.

KIRKLIN, J. K., WATSON, M., BONDOC, C. C. and BURKE, J. F. (1976), Treatment of hydrazine-induced coma with pyridoxine, *New England Journal of Medicine*, **294**, 938-9.

KROP, S. (1954), Toxicology of hydrazine, *Archives of Industrial Hygiene and Occupational Medicine*, **9**, 199-204.

LEVIN, P. J., KLAFF, L. J., ROSE, A. G., and FERGUSON, A. D. (1979), Pulmonary effects of contact exposure to paraquat: a clinical and experimental study, *Thorax*, **34**, 150-60.

LEAVITT, J. R. C., GOLD, R. E., HOLCSLAW, T. and TUPY, D. (1982), Exposure of professional pesticide applicators to carbaryl, *Archives of Environmental Contamination and Toxicology*, **11**, 57-62.

MCCONNELL, E. E., BUCHER, J. R., SCHWETZ, B. A., GUPTA, B. N., SHELBY, M. D., LUSTER, M.I., BRODY, A. R., BOORMAN, G. A., RICHTER, C, STEVENS, M. A. and ADKINS, B. JR., (1987), Toxicity of methyl isocyanate, *Environmental Science and Technology*, **21**, 188-93.

NEWHOUSE, M., MCEVOY, D., and ROSENTHAL, D. (1978), Percutaneous paraquat absorption, *Archives of Dermatology*, **114**, 1516-9.

NIOSH (1978), *Criteria for a recommended standard, occupational exposure to hydrazines*, (Cincinnati, OH: National Institute for Occupational Safety and Health).

O'BRIEN, R. D. and DANELLEY, C. E. (1965), Penetration of insecticides through rat skin, *Journal of Agricultural Food Chemistry*, **13**, 245-247.

OKONEK, S., WRONSKI, R. NIEDERMAYER, W., OKONEK, M. and LAMER, A. (1983), Near fatal percutaneous paraquat poisoning, *Klinische Wochenschrift*, **61**, 655-9.

PASI, A. (1978), *The toxicology of paraquat, diquat and morfamquat* (Bern: Hans Huber).

PETERSEN, P., BREDAHL, E., LAURITSEN, O. and LAURSEN, T. (1970). Examination of the liver in personnel working with liquid rocket propellant, *British Journal of Industrial Medicine*, **27**, 141-6.

PEVNY, I. and PETER, G. (1983), Allergisches Kontaktexzem aus Pyridin- und Hydrazinderivate, *Dermatosen in Beruf und Umwelt* **31**, 78–83.

REIDENBERG, M. M., DURANT, P. J., HARRIS, R. A., DE BOCCARDO, G., LAHITA, R. and STENZEL, K. H. (1983), Lupus erythematosus-like disease due to hydrazine, *American Journal of Medicine*, **75**, 365–70.

ROSENBERG, C. and SAVOLAINEN, H. (1986), Determination of occupational exposure to toluene diisocyanate by biological monitoring, *Journal of Chromatography*, **367**, 385–92.

ROTHE, A (1976), Occupational skin disease caused by polyurethane chemicals (in German), *Berufsdermatosen*, **24**, 7–24.

ROTHE, A. (1988), Contact dermatitis from *N*-(*o*-chlorobenzylidene)phenylhydrazine, *Contact Dermatitis*, **18**, 16–9.

SCHUCKMANN, F. (1969), Observations on different forms of phenlylhydrazine intoxication, *Zentralblatt für Arbeitsmedizin und Arbeitsschutz*, **19**, 388–91.

SHAH, P. V. and GUTHRIE, F. E. (1977), Dermal absorption, distribution and the fate of six pesticides in the rabbit, In *Pesticide Management and Insecticide Resistance*, eds. D.L. Watson and A. W. A. Brown, pp. 547–554 (New York: Academic Press).

SHAH, P. V. and GUTHRIE, F. E. (1983), Percutaneous penetration of three insecticides in rats: A comparison of two methods for in vivo determination, *Journal of Investigative Dermatology*, **80**, 291–293.

SHAH, P. V., MONROE, R. J. and GUTHRIE, F.E. (1981), Comparative rates of dermal penetration of insecticides in mice, *Toxicology and Applied Pharmacology*, **59**, 414–423.

SHAH, P.V., FISHER, H.L., SUMLER, M.R., MONROE, R.J., CHERNOFF, N. and HALL, L. L., (1987), Comparison of the penetration of 14 pesticides through the skin of young and adult rats, *Journal of Toxicology and Environmental Health*, **21**, 252–66.

SHOOK, B. S. Sr. and COWART, O. H. (1957) Health hazards associated with unsymmetrical dimethylhydrazine, *Industrial Medicine and Surgery*, **26**, 333–6.

SMITH, E. B. and CLARK, D. A. (1971), Absorption of unsymmetrical dimethylhydrazine (UMDH) through canine skin, *Toxicology and Applied Pharmacology*, **18**, 649–59.

SMITH, E. B. and CLARK, D. A. (1972), Absorption of hydrazine through canine skin, *Toxicology and Applied Pharmacology*, **21**, 189–93.

SMYTH, H. F. and CARPENTER, C. P. (1948), Further experience with the range finding test in the industrial toxicology laboratory, *Journal of Industrial Hygiene and Toxicology*, **30**, 63–8.

SMYTH, H. F., CARPENTER, C. P., WEIL C. S. and POZZANI, U. C. (1954), Range-finding toxicity data: List V. *Archives of Industrial Hygiene and Occupational Medicine* **10**, 61–8.

SMYTH, H. F., CARPENTER, C. P., WEIL, C. S., POZZANI, U. C., STRIEGEL, J. A. and NYCUM, J. S. (1969), Range-finding toxicity data: List VII, *American Industrial Hygiene Association Journal*, **30**, 470–476.

SOLOMONS, B. (1946), A case of allergy to phenylhydrazine, *British Journal of Dermatology*, **58**, 236–7.

SOTANIEMI, E., HIRVONEN, J., ISOMÄKI, H., TAKKUNEN, J. and KAILA, J. (1971) Hydrazine toxicity in the human, report of a fatal case, *Annals of Clinical Research*, **3**, 30–3.

SWAN, A.A.B. (1969), Exposure of spray operators to paraquat, *British Journal of Industrial Medicine*, **26**, 322–9.

THIESS, A. M. (1965), Gesundheitsschädigungen und Vergiftungen durch Einwirkung von Äthylenimin, *Archives of Toxicology*, **21**, 67–82.

TUNGSANGA, K., CHUSILP, S., IRASENA, S., and SITPRIJA, V. (1983), Paraquat poisoning: evidence of systemic toxicity after dermal exposure, *Postgraduate Medical Journal* **59**, 338–9.

VALE, J. A., MEREDITH, T. J. and BUCKLEY, B. M. (1987), Paraquat poisoning: clinical features and immediate general management, *Human Toxicology* **6**, 41-7.

VERNOT, E. H., MACEWEN, J. D., HAUN, C. C. and KINKEAH, E. R. (1977), Acute toxicity and skin corrosion data for some organic and inorganic compounds and aqueous solutions, *Toxicology and Applied Pharmacology*, **42**, 417-23.

WALKER, M., DUGARD, P. H. and SCOTT, R. C. (1983), Absorption through human and animal skins: in vitro comparisons, *Acta Pharmaceutica Suecica*, **20**, 52-3.

WESTER, R. C., MAIBACH, H. I., BUCKS, D. A. W. and AUFRERE, M. B. (1984), In vivo percutaneous absorption of paraquat from hand, leg, and forearm of humans, *Journal of Toxicology and Environmental Health*, **14**, 759-62.

WHEELER, C. E. JR., PENN, S. R. and CAWLEY, E. P. (1965), Dermatitis from hydrazine hydrobromide solder flux, *Archives of Dermatology*, **91**, 235-9.

WHO (1982) *Recommended health-based limits in occupational exposure to pesticides.* Report of a WHO Study Group (Technical Report Series No. 677) (Geneva: World Health Organization).

WHO (1984), *Paraquat and diquat*, (Environmental Health Criteria 39) (Geneva: World Health Organization).

WHO (1987), *Hydrazine*, (Environmental Health Criteria 68) (Geneva: World Health Organization).

WILSON D. J. B. (1930), Nicotine poisoning by absorption through the skin, *British Medical Journal*, **2**, 601-2.

WOHLFAHRT, D. J. (1982), Fatal paraquat poisonings after skin absorption, *Medical Journal of Australia* **1**, 512-3.

YAKIM, V. S. (1967), The maximum permissible concentration of Sevin in the air of the work zone (in Russian), *Gigiena i Sanitariia*, **32**(4), 29-33.

ZWEIG, G., GAO, R.-Y., WITT, J. M., POPENDORF, W. J. and BOGEN, K. T. (1985), Exposure of strawberry harvesters to carbaryl. In *Dermal exposure related to pesticide use*, (ACS symposium series 273), eds. R.C. Honeycutt, G. Zweig and N. N. Ragsdale, pp. 123-38. Washington, DC: American Chemical society).

Chapter 11
Heterocyclic oxygen and sulfur compounds

Introduction

The heterocyclic compounds most frequently considered to be skin absorption hazards are indicated in table 11.1. The most important of these are considered separately below.

Most of these compounds are reactive and can cause severe adverse effects upon contact with the skin. Limited information on percutaneous absorption is available. However, the caustic effects on the skin may perhaps promote the percutaneous transfer of the compound to the systemic circulation. In such cases, toxicity may develop in other parts of the body.

In addition to those listed in the table, other heterocyclic compounds with a 'skin' denotation in a single country include ethylene oxide, a probable carcinogen (class 2A) (IARC, 1987), and *beta*-propiolactone (2-oxetanone), a class 2B carcinogen. Ethylene sulfide, and tetrahydrofuran (butylene oxide) are also listed in one country only.

Warfarin, a common rodenticide, is considered a skin hazard in one country.

Table 11.1. Heterocyclic oxygen and sulfur compounds that are considered a skin hazard in several countries (y = yes, n = no, o = other regulation). Those that are described in separate sections are italicized.

CAS No.	Chemical	Number of countries	FRG	Sweden	USA
106-89-8	*Epichlorohydrin*	12	y	y	y
106-92-3	Allyl glycidyl ether (allyl 2,3-epoxy -1-propyl ether)	8	o	n	y
98-00-0	Furfuryl alcohol	6	o	o	y
98-01-1	Furfural	12	y	y	y
123-91-1	*1,4, Dioxane*	13	y	y	y

Epichlorohydrin

Skin exposure potential

Epichlorohydrin (3-chloro-1,2-epoxypropane) is mainly used in the manufacture of glycerine and unmodified epoxy resins, but it is also used for the production of elastomers, surface active agents, pharmaceuticals, insecticides and solvents. Within the EC the annual production in 1985 was 170 000 tonnes, and 100 000 tonnes were consumed.

Physicochemical properties

Epichlorohydrin is a liquid at room temperature, the boiling point is 116°C, and the vapour pressure at 20°C is 13 mm Hg. Epichlorohydrin is miscible with water (6.4% at 20°C), ethanol, diethyl ether and chlorinated aliphatic hydrocarbon solvents, but immiscible with petroleum hydrocarbons. The solubility characteristics of this compound are not very different from those of psoriasis scales (Hansen and Andersen, 1988), and a considerable uptake in the skin could therefore be expected. The small molecular size of this non-polar compound would suggest that it is capable of rapidly penetrating the skin, although evaporation and chemical reactions in the skin would limit the total amount absorbed.

Experimental data

The dermal LD_{50} for the rabbit is 755 mg kg^{-1}, i.e. less than five-fold more than that due to intraperitoneal injection (Lawrence et al., 1972). Seven out of 10 mice died, mostly within 24 h, after a 15–20 min immersion of the tail in epichlorohydrin; the survivors developed necrotic lesions on the tail; a main sign of toxicity was nephrotoxicity with anuria (Pallade et al., 1967). Application of 0.5 ml of epichlorohydrin to rabbit skin resulted in lesions with a central necrotic zone within 24 h, while 0.1–0.2 ml applied for 2 h caused less severe lesions (Pallade et al., 1967). The systemic toxicity leads to CNS depression with muscular paralysis and depressed respiration which ultimately leads to death of the animals. The allergenic potential of epichlorohydrin and epoxy resins has been documented in experimental studies (Thorgeirsson and Fregert, 1977). Sufficient evidence is available for carcinogenicity in animals (IARC, 1987).

Human data

Skin contact is followed by a delayed and painful irritation of the subcutaneous tissues (Ippen and Mathies, 1970). Epichlorohydrin is the main component responsible for contact dermatitis due to epoxy resins (Lambert et al., 1978; Beck and King, 1983; Prens et al., 1986). Limited evidence is available on systemic toxicity in humans and the contribution by skin absorption to adverse effects. Inadequate evidence is available to evaluate the possible carcinogenicity in humans (Tassignon et al., 1983; WHO, 1984) and it is considered a class 2A carcinogen (IARC, 1987).

Conclusions

Epichlorohydrin is a reactive compound which may penetrate the skin and cause serious local damage and sensitization. Experimental studies document that systemic toxicity may be caused by skin exposure, but data from human studies are unavailable. Epichlorohydrin is regarded a probable carcinogen. Epichlorohydrin should therefore be regarded as a skin absorption hazard.

1,4-Dioxane

Skin exposure potential

1,4-Dioxane (p-dioxane) is a widely used solvent, in particular in fibre production from cellulose esters and ethers. Also, a considerable amount is used in laboratories and in pharmaceutical and cosmetic products. Although accurate figures are not available, only a few thousand individuals were thought to be exposed to dioxane in western Europe, and only about 10 at production plants (ECETOC, 1983).

Physicochemical properties

Dioxane is a liquid at room temperature with a vapour pressure of about 40 mm Hg. Dioxane is miscible with water and most organic solvents (NIOSH, 1977). The partition coefficient between toluene and water was reported to be about 1.4 (Fairley et al., 1934). The log P_{ow} is negative, i.e -0.42, suggesting a high degree of hydrophilicity. The solubility characteristics are not very different from those of psoriasis scales (Hansen and Andersen, 1988), and a considerable uptake in the skin could therefore be expected. With a small molecular size, this compound would be capable of rapidly penetrating the skin. However, evaporation from the skin and oxidation would limit the total amount absorbed.

Experimental data

Percutaneous absorption of ^{14}C-labelled dioxane was studied by Marzulli et al. (1981). From a methanol solution and a skin lotion, 2.3 and 3.4% of the dioxane was absorbed, and the highest rate of absorption was within the first 4 h. This study documents that dioxane may be absorbed through the skin both when present in polar and non-polar vehicles; the total amount absorbed will be limited by the volatility of the compound, unless occlusion of the exposed skin is present. Experimental studies by Fairley et al. (1934) and Nelson, (1951) show that severe or fatal dioxane poisoning may be induced by skin exposure. A dermal LD_{50} in rabbits of 7600 mg kg^{-1} (RTECS, 1983) is relatively high, though not much higher than LD$_{50}$ values obtained for dioxane using other administration routes. Toxic effects mainly occur in the liver, kidney and central nervous system. Skin painting with dioxane resulted in promotion of tumour development caused by dimethylbenzanthracene (King et al., 1973). Sufficient evidence is available for carcinogenicity to animals (IARC, 1987).

Human data

In one fatal dioxane intoxication, a worker had used this solvent extensively to keep his hand free of glue during one week; respiratory dioxane exposure occurred, but skin absorption may have caused a significant contribution (Johnstone, 1959). Toxic effects on central nervous system, liver and kidney have been seen at autopsies, and the systemic toxicity of dioxane has been documented in several epidemiological studies (NIOSH, 1977). The contribution to total absorption levels by percutaneous penetration has not been documented in these reports. Skin irritation may occur, and one case of allergic contact dermatitis has been reported (Fregert, 1974). However, current industrial hygiene practices seem to limit considerably the total exposures to dioxane (Thiess et al., 1976). Dioxane is regarded as a possible carcinogen, i.e. class 2B (IARC, 1987).

Conclusions

Dioxane may rapidly penetrate the skin, although the high volatility will limit the amount absorbed, unless occlusion occurs. Due to the systemic toxicity and potential carcinogenicity, dioxane should be regarded as a skin exposure hazard.

Other compounds

Allyl glycidyl ether ((2-propenyloxy)methyloxirane or allyl 2,3-epoxy-1-propyl ether) has structural resemblence to epichlorohydrin. This and other epoxy compounds are widely used, and dermal exposure appears to be significant. It is a primary irritant and is suspected of being a carcinogen.

Furfuryl alcohol is used in adhesives and is also a primary irritant. The dermal LD$_{50}$ in rabbits is 400 mg kg^{-1} (Deichmann, 1969). It has shown mutagenic effects. Little information is available on the systemic significance of cutaneous exposure to this compound.

Furfural is mainly a chemical intermediate. It is also a primary irritant with mutagenic effects. Assuming no toxic effect of the compound on the stratum corneum, a penetration rate of 1.82 mg cm^{-2} h^{-1} was calculated on the basis of physicochemical properties; this rate would mean that exposure of a skin area of 360 cm^2 to furfural would cause an absorption 300-fold above the uptake from inhalation of furfural at the TLV level for the same time period (Fiserova-Bergerova and Pierce, 1989). However, this calculation is slightly misleading, because the main purpose of the exposure limit is to protect against respiratory irritation, rather than systemic toxicity. A study of urinary excretion of urinary metabolites in humans indicated an approximate percutaneous absorption rate of $0.2\,\text{mg cm}^{-2}\text{h}^{-1}$, a 15-min exposure of one hand to furfural would then lead to an absorption similar to the one following inhalation of $10\,\text{mg m}^{-3}$ for 8 h (Flek and Sedivec, 1978).

References

BECK, M. H., and KING, C. M. (1983), Allergic contact dermatitis to epichlorohydrin in a solvent cement, *Contact Dermatitis*, **9**, 315.

DEICHMANN, W. B., ed. (1969), *Toxicology of drugs and chemicals*, p. 280 (New York: Academic Press).

ECETOC (1983), *Joint assessment of commodity chemicals, No. 2, 1,4-dioxane*, (Brussels: European Chemical Industry Ecology & Toxicology Centre).

FAIRLEY, A., LINTON, E. C., and FORD-MOORE, A. H. (1934), The toxicity to animals of 1,4-dioxane, *Journal of Hygiene*, **34**, 486–501.

FISEROVA-BERGEROVA, V. and PIERCE, J. T., (1989), Biological monitoring V. Dermal absorption, *Applied Industrial Hygiene*, **4**, F14-F21.

FLEK, J. and SEDIVEC, V. (1978), The absorption, metabolism and excretion of furfural in man, *International Archives of Occupational and Environmental Health*, **41**, 159–68.

FREGERT, S. (1974), Allergic contact dermatitis from dioxane is a solvent for cleaning metal parts, *Contact Dermatitis Newsletter* **15**, 438.

HANSEN, C. M. and ANDERSEN, B. H. (1988). The affinities of organic solvents in biological systems, *American Industrial Hygiene Association Journal*, **49**, 301–8.

IARC (1987), *Overall evaluations of carcinogenicity: An updating of IARC Monographs Volumes 1 to 42*, (IARC Monographs on the evaluation of carcinogenic risk of chemicals to man (Suppl. 7.) (Lyon: International Agency for Research on Cancer).

IPPEN, H., and MATHIES, V. (1970), Die "protrahierte Verätzung", *Berufsdermatosen* **18**, 144–65.

JOHNSTONE, R. T. (1959), Death due to dioxane? *American Medical Association, Archives of Industrial Health*, **20**, 445–7.

KING, M. E., SHEFNER, A. M., and BATES, R. R. (1973), Carcinogenesis bioassay of chlorinated dibenzodioxines and related chemicals, *Environmental Health Prospective*, **5**, 163–70.

LAMBERT, D., LACROIX, M., DUCOMBS, G., JOURNET, F. and CHAPUIS, J.-L. (1978), L'allergie cutanée a l'épichlorhydrine, *Annals of Dermatology and Venereology*, **105**, 521–5.

LAWRENCE, W. H., MALIK, M., TURNER, J. E., and AUTIAN, J. (1972), Toxicity profile of epichlorohydrin, *Journal of Pharmaceutical Science*, **61**, 1712–7.

MARZULLI, F. N., ANJO, D. M., and MAIBACH, H. I., (1981), In vivo skin penetration studies of 2,4-toluenediamine, 2,4-diaminoanisole, 2-nitro-p-phenolenediamine, p-dioxane and N-nitrosodiethanolamine in cosmetics, *Food and Cosmetics Toxicology* **19**, 743–7.

NELSON, N. (1951), Solvent toxicity with particular reference to certain octyl alcohols and dioxanes, *Medical Bulletin*, **11**, 226–38.

NIOSH (1977), *Criteria for a recommended standard... Occupational exposure to dioxane*, (DHEW (NIOSH) Publication No. 77-226) (Cincinnati: National Institute for Occupational Safety and Health).

PALLADE, S., DOROBANTU, M., ROTARU, G. and GABRILESCU, E. (1967), Étude expérimentale de l'intoxication par l'épichlorhydrine, *Archives des maladies professionnelles de médicine du travail et de sécurité sociale*, **28**, 505-16.

PRENS, E.P., DE JONG, G., and VAN JOOST, T. (1986), Sensitization to epichlorohydrin and epoxy system components, *Contact Dermatitis*, **15**, 85-90.

RTECS (1983), *Registry of toxic effects of chemical substances*, (Suppl). (Cincinnati Ohio: National Institute for Occupational Safety and Health).

TASSIGNON, J.P., BOS, G.D., CRAIGEN, A.A., JACQUET, B., KUENG, H.L., LANOUZIERE-SIMON, C., and PIERRE, C. (1983), Mortality in a European cohort occupationally exposed to epichlorohydrin (ECH), *International Archives of Occupational and Environmental Health*, **51**, 325-36.

THIESS, A.M., TRESS, E. and FLEIG, I. (1976), Arbeitsmedizinische Untersuchungsergebnisse von Dioxan-exponierten Mitarbeitern, *Arbeitsmedizin, Sozialmedizin und Präventivmedizin*, **11**, 36-46.

THORGEIRSSON, A., and FREGERT, S. (1977), Allergenicity of epoxy resins in the guinea pig, *Acta Dermatovener*, **57**, 253-6.

WHO (1984), *Epichlorohydrin (Environmental health criteria 33)* (Geneva: World Health Organization).

Chapter 12
Organophosphorus compounds

Introduction

The organophosphorus compounds most frequently considered to be skin absorption hazards are indicated in table 12.1. The most important of these are considered separately below.

Serious skin exposures have occurred in relation to application of organophosphorus pesticides; drenched clothes, lack of protective equipment and unsafe spraying methods have resulted in thousands of fatal intoxications, mostly due to percutaneous absorption (WHO, 1982). The organophosphorus pesticides inhibit cholinesterase, and this effect results in blurred vision, headache, nausea and vomiting; in more severe cases, profuse sweating, pinpoint pupils, salivation, muscular weakness, tremor and difficulty in breathing may follow (WHO, 1986).

Although less extensive information is available on organophosphorus pesticides other than those italicized in the table, absorption through the skin is probably a significant hazard. These compounds share a low vapour pressure,

Table 12.1. Organophosphate compounds that are considered a skin hazard in several countries (y = yes, n = no, o = other regulation). Those that are considered in detail are italicized.

CAS No.	Chemical	Number of countries	FRG	Sweden	USA
Aliphatic phosphates					
107–49-3	TEPP (tetraethyl pyrophosphate)	7	y	n	y
3689-24-5	Sulfotep (tetraethyl pyrothiphosphate)	8	y	n	y
298-02-2	Phorate (Thimet)	6	o	n	y
8065-48-3	Demeton	9	y	n	y
8022-00-2	Demeton-methyl	9	y	n	y
62-73-7	*Dichlorvos*	11	y	n	y
7786-34-7	Mevinphos (Phosdrin)	9	y	n	y
121-75-5	*Malathion*	10	o	n	y
563-12-2	Ethion	6	o	n	y
Isocyclic phosphates					
2104-64-5	EPN (*O*-ethyl-*O*-(4-nitrophenyl)phenyl-phosphonothioate)	10	y	n	y
298-00-0	Methyl parathion	11	o	n	y
56-38-2	*Parathion*	14	y	n	y
Heterocyclic phosphates					
37280-01-6	*Diazinon*	9	y	n	y
86-50-0	Azinphos-methyl (Guthion)	8	y	n	y

slight solubility in water and a relatively large molecular size which decreases but does not prohibit cutaneous penetration. Percutaneous penetation tends to increase with the degree of water solubility, and the penetration rate depends on the solvent used (Dedek, 1980).

Other aliphatic organophosphates with skin denotation in some countries are dicrotophos, disulfoton, schradan, butiphos, and hexamethylphosphoric acid triamide. Additional isocyclic phosphates are fonofos, fenthion, fenitrothion and fenamiphos. Chlorpyrifos is a heterocyclic phosphate with a 'skin' denotation in five countries.

Tri-*o*-cresyl phosphate has a 'skin' denotation in three countries. This compound is not a pesticide; it has caused outbreaks of peripheral neuropathy.

Dichlorvos

Skin exposure potential

Dichlorvos (*O,O*-dimethyl-*O*-2,2,-dichlorovinyl phosphate, DDVP) is an organophosphorus pesticide. The annual production in EC countries was 8–10 thousand tonnes in 1985, with most of that being consumed within the EC. Dichlorvos is used, e.g. for control of external parasites on livestock, insects in buildings and flea collars for pets. Cutaneous exposure of some pest control operators inside a building was 0.05–0.08 μg cm^{-2} by the end of the workday (Das *et al.*, 1983) and for dichlorvos applicators in an orchard 0.7 μg cm^{-2} h^{-1} (IARC, 1979).

Physicochemical properties

Dichlorvos is a liquid at room temperature. The vapour pressure is 1.2×10^{-2} mm Hg at 20°C. It is slightly soluble in water (9 g l^{-1}) and miscible with most organic solvents. The octanol/water partition coefficient (log P_{ow}) of 3.47, thus indicating considerable lipophilicity. With a molecular mass of 221, dichlorvos would be expected to penetrate the skin barrier at a relatively slow rate.

Experimental data

Dermal LD$_{50}$ values of 107 and 75 mg kg^{-1} in male and female rats, respectively, as compared to oral LD$_{50}$ values of 80 and 56 mg kg^{-1} (Gaines, 1969), would suggest a considerable skin penetration potential. However, other studies have shown larger differences between the two routes of absorption (WHO, 1989). Dermal LD$_{50}$ values depend on the solvent used and the presence of occlusion (Jones *et al.*, 1968). Cutaneous application of dichlorvos to rabbits produced a measurable inhibition of cholinesterase as an indication of systemic toxicity (Shellenberger, 1980), but quantitative data were not given. A dermal dose of 50 mg kg^{-1} produced cholinergic signs in a monkey 20 min after administration;

after eight daily doses at this level, the animal died (WHO, 1978). Daily dermal applications of dichlorvos at 21.4 mg kg^{-2} to male rats for 5 days a week for 13 weeks did not result in any discernible toxicity (Dikshith et al., 1976), probably due to the volatility of the compound; cholinesterase was not measured in this study.

Human data

Several fatal and non-fatal cases of dichlorvos intoxication have been described after concentrated formulations were splashed onto parts of the body; fatalities were seen in cases where the pesticide was not promptly washed off (WHO, 1989). Leaking of dichlorvos onto the shoulder of a pesticide applicator resulted in dizziness and rapid, shallow breathing; 3 days later the blood cholinesterase level was 36% of normal (Biesby and Simpson, 1975). A similar, but less severe case was reported by Mathias (1983). Dichlorvos may cause allergic contact dermatitis (Matsushita et al., 1985). Systemic toxicity is due to cholinesterase inhibition; two cases of acute dichlorvos intoxication after oral intake later resulted in axonal degeneration neuropathy (WHO, 1989).

Conclusions

Available evidence suggests that dichlorvos may be absorbed through the skin, in particular when evaporation is limited by occlusion. Dichlorvos causes toxicity due to inhibition of cholinesterase and should therefore be regarded as a skin absorption hazard.

Malathion

Skin exposure potential

Annual production of malathion (S-1,2-bis(ethoxycarbonyl)ethyl-O,O-dimethyl phosphorodithioate) within the EC was 7000 tonnes in 1979. Malathion is marketed as water-dispersible powders, emulsifiable concentrates, and dust formulations. It is used for controlling pests affecting various crops, such as cereals, beans, grapes and citrus fruits. Malathion also has veterinary uses. Occupational exposures occur in the production, formulation and application of malathion. Dermal exposure to malathion in pesticide sprayers may considerably exceed the inhalation exposures and result in significant inhibitions of cholinesterase (Baker et al., 1978). Skin and inhalation exposures of malathion sprayers on tractors were compared by skin-patch testing and air sampling: airborne levels were quite low and averaged about 0.6 mg m^{-3}, while the dermal exposure was estimated at 2.5 mg h^{-1} (Jegier, 1974). A recent study used a fluorescent tracer and documented that dermal exposures in air-blast applicators were higher than in mixers and correlated closely with subsequent malathion metabolite excretion (Fenske, 1988). Exposure conditions vary considerably with work practices and other studies have indicated that improved hygiene would not result in detectable toxicity (WHO, 1982).

Physicochemical properties

Technical malathion is a colourless or yellowish liquid at room temperature, but the purified compound is a crystalline solid. The vapour pressure at 20°C is 4×10^{-5} mm Hg, and the boiling point of the technical product is about 156°C. Malathion is moderately soluble in water, and 1 l will dissolve 145 mg at 25°C. The technical product is miscible with most organic solvents, but the solubility in petroleum oils is limited. The solvent/water partition coefficient for chloroform and benzene is about 100. The octanol/water partition coefficient (log P_{ow}) is 2.89, thus indicating a lipophilic character of the compound. With a molecular weight of 330, malathion would be expected to permeate skin relatively slowly.

Experimental data

The oral LD_{50} of malathion was 1375 and 1000 mg kg^{-1} for male and female rats, and the dermal LD_{50} values were above 4444 mg kg^{-1} (Gaines, 1969). Additional information reviewed by a WHO Study Group (1982) suggested that dermal application results in LD_{50} values only slightly above those resulting from oral intake. When applied in an acetone solution to the shaved skin of mice, half of the malathion was found to have penetrated the skin by 130 min, i.e. a relatively slow absorption rate (Shah et al., 1981). Skin penetration of malathion has also been demonstrated in rats where the half-time for penetration was 5.5 h (O'Brien and Danelley, 1965) and in rabbits where the peak blood level was reached about 3 h after cutaneous application (Shah and Guthrie, 1977). Experimental animal studies have shown a wide variability of malathion toxicity, to a large degree probably related to the contents of impurities (Aldridge et al., 1979). Unfortunately, detailed dose–response information is unavailable, because many malathion preparations contain toxic impurities that have only been determined in the most recent studies.

Human data

Absorption in humans has been studied by the indirect method where the excretion of radioactivity in the urine is followed after a dermal application of an acetone solution of radioactively labelled malathion to the skin of the forearm: an average of about 8% was absorbed in six subjects during 5 days; although the half-life after intravenous injection appeared to be only 3 h, a maximal excretion in the urine was seen about 8 h after dermal application (Feldman and Maibach, 1974). In a study of 39 male subjects dusted with talcum powder containing a total of 90 g of malathion over 8–16 weeks, the urinary excretion suggested that about 3% of the malathion available for absorption actually penetrated the skin (Hayes et al., 1960). The permeability of the skin varies: application on the forearm, palm of the hand and ball of the foot resulted in a 7% absorption, somewhat higher absorption was seen on abdomen and the dorsum of the hand, and three- to four-fold increases were seen in the skin of the forehead and the axilla (Maibach et al., 1971). A recent study has shown that repeated skin application of malathion does not change the absorption rate significantly (Wester et al., 1983), and results from single-dose studies would therefore seem to be reliable for prediction of long-term absorption. One grade of malathion has been reported to act as a skin sensitizer (Milby and Epstein, 1964). The information collected by a WHO Study Group (1982) suggests that occupational malathion intoxication is primarily caused by skin absorption. Signs and symptoms of malathion intoxication are due to the inhibition of cholinesterase (WHO, 1982). Several hundred cases of malathion poisoning have occurred, but the fatal cases have all been due to accidental or suicidal ingestion of large quantities of the pesticide (WHO, 1982). Delayed effects have not been documented.

Conclusion

Malathion is capable of passing through the skin, and dermal exposures have been convincingly implicated as causes of human intoxications from malathion.

EPN

Skin exposure potential

EPN (*O*-ethyl-*O*-(4-nitrophenyl)phenylphosphonothioate is an organophosphorus pesticide. It is particularly used on cotton fields. Mean 8-h dermal exposures for ground applicators was 7.5 mg, somewhat less in pilots and loaders, but 118 mg for flagmen; respiratory exposures were much less (Atallah et al., 1982).

Physicochemical properties

EPN is a solid at room temperature with a melting point of 36°C. Thus, at room temperature, the vapour pressure is minimal. It is structurally related to parathion. EPN is only slightly soluble in water, but soluble in acetone, other ketones and benzene. The octanol/water partition coefficient (log P_{ow}) is 1.70. The lipophilicity and the low vapour pressure would make a prolonged percutaneous absorption possible.

Experimental data

In rats, the oral LD_{50} is 36 and 7.7 mg kg^{-1} for male and female animals, and the dermal LD_{50} is 230 and 25 mg kg^{-1}, respectively (Gaines, 1969). Repeated dermal applications of EPN result in a longer elimination half-life compared to a single application (Abou-Donia et al., 1983). Acute toxicity is due to inhibition of cholinesterase with resulting accumulation of acetylcholine; the major signs include sweating, salivation, diarrhoea, bradycardia, bronchoconstriction, muscle fasciculations and coma. In chickens, EPN has produced a delayed demyelination syndrome; histopathological changes were seen at 0.1 mg kg^{-1} day^{-1} (Abou-Donia, 1981). Clinical effects (ataxia) were produced by dermal application of 1.3–1.5 mg kg^{-1} day^{-1} for 50–80 days in hens (Francis et al., 1985).

Human data

After skin absorption, localized sweating and muscular fasciculations occur in the immediate area, usually within 15 min to 4 h; skin absorption is greater at higher temperatures and is increased by the presence of dermatitis (NIOSH/OSHA, 1978). Systemic toxicity is due to the inhibition of cholinesterase. One case of delayed neuropathy was reportedly associated with exposure to EPN (Petty, 1958), but the potential of EPN of producing demyelination in humans is still unclear.

Conclusions

Limited evidence suggests that EPN can be absorbed through the skin and that systemic toxicity may ensue. EPN should therefore be regarded as a skin absorption hazard.

Parathion

Skin exposure potential

Parathion (O,O-diethyl-O-(4-nitrophenyl)phosphorothioate) is an organophosphorus pesticide that has been known for more than 40 years. It is mainly insecticidal and acaricidal, but its considerable mammalian toxicity has resulted in decreased use; in some countries, parathion is banned. Skin and inhalation exposures of parathion sprayers on tractors were compared by skin-patch testing and air sampling: airborne levels were quite low and averaged about 0.15 mg m^{-3}, while dermal exposure was estimated at a much higher level of 2.4 mg h^{-1} (Jegier, 1974). The total production within the EC was 6000 tonnes in 1984, but most was exported, and consumption within the community was less than 1000 tonnes (ECDIN, pers. comm. 1987). Thus, the skin exposure potential is probably decreasing.

Physicochemical properties

Parathion is a yellowish liquid at room temperature. The melting point is 6°C, and the vapour pressure at 20°C is only 0.003 mm Hg. Parathion is sparingly soluble in water (24 mg l^{-1}), but is miscible with ethanol, benzene and several other solvents, except for paraffinic solvents in which parathion is only slightly soluble (NIOSH, 1976). Octanol/water partition coefficients (log P_{ow}) for parathion products have been reported to be 2.15 and 3.81, thus indicating a high degree of lipophilicity. With a molecular weight of 291, parathion would be expected to penetrate the skin only slowly. However, the low vapour pressure and the solubility in stratum corneum would make a prolonged absorption possible. A penetration rate of 0.015 mg cm^{-2} h^{-1} was calculated on the basis of physicochemical properties (Fiserova-Bergerova and Pierce, 1989).

Experimental data

The dermal LD$_{50}$ has been found to be 21 mg kg^{-1} in male rats and somewhat less in female rats (Gaines, 1969). In guinea pigs, the dermal LD$_{50}$ was about 1 mg kg^{-1} (Roudabush et al., 1965), and the results given for rabbits vary considerably. In general, the dermal LD$_{50}$ values are not much above those seen with other administration routes. *In vitro* studies of parathion penetration of human skin indicated a low flux of 0.6 μg cm^{-2} h^{-1} (Frederiksson, 1961). A slightly higher rate was obtained in a study using rabbits, but the absorption rate varied considerably (Nabb et al., 1966). When applied in an acetone solution to the shaved skin of mice, half of the ^{14}C-parathion was found to have penetrated the skin by about 1 h, i.e. an absorption rate somewhat faster than that for malathion (Shah et al., 1981); ^{14}C-parathion applied in an acetone solution to 2 cm^2 of shaved skin of rabbits also penetrated faster than did malathion (Shah and Guthrie, 1977). An absorption rate of 0.33 and 0.49 μg cm^{-2} h^{-1} was found for adult male and female rats, respectively; the permeability constants were 7.5×10^{-3} and 1.0×10^{-2} cm h^{-1} (Knaak et al., 1984). Thus, following some evaporation after application, most of the parathion applied to the skin is absorbed within 3–5 days (Knaak et al., 1984; Shah et al., 1987). That parathion absorbed by the dermal route appears less toxic than by inhalation or ingestion may be due to metabolism in the cutaneous tissues (Holmstedt, 1959; Frederiksson et al., 1961). However, such partial breakdown in no way eliminates the hazard. In addition to the slow absorption through the skin, further delay is caused by the slow formation of paraoxon; this toxic metabolite is responsible for inhibition of cholinesterase and the resulting clinical effects (NIOSH, 1976). One study showed that monkeys, after application of 20–80 mg of parathion to the skin, continued to excrete a parathion metabolite for 18–30 days (IRPTC, 1982).

Human data

After dermal application of an acetone solution of ^{14}C-parathion to the skin of the forearm of human volunteers, urinary excretion of radioactivity suggested an average absorption of almost 10% within 5 days, i.e. a slightly higher accumulated absorption than seen with malathion (Feldman and Maibach, 1974). Considerable regional variation in percutaneous parathion absorption has been documented (Maibach et al., 1971). Dermal absorption may be increased if xylene, a known skin irritant, is used as a vehicle (NIOSH, 1976). In parathion application, skin exposure is a more serious hazard than is respiratory exposure, and a parathion metabolite can be detected in urine up to 10 days after the last exposure (Durham et al., 1972). Percutaneous absorption of parathion has given rise to numerous cases of poisoning (NIOSH, 1976; Jia-huao et al., 1985). Following parathion spraying, serious percutaneous uptake may occur several days later as a result of skin contact with the sprayed fruits and leaves; as parathion causes no skin irritation and the absorption is slow, serious poisoning may develop unexpectedly after the worker has returned home (NIOSH, 1976). Due to the slow regeneration of cholinesterase activity, continued dermal absorption of parathion over several days may lead to cumulative toxicity. The neurotoxic effects are mainly muscarinic and nicotinic with gastroenteritis, sweating, salivation, lacrimation, bradycardia, miosis, blurred vision and muscular fasciculations (NIOSH, 1976).

Conclusions

Parathion penetrates the skin only slowly, but its low vapour pressure and lipophilicity may lead to a protracted uptake and cumulated toxicity. Both case reports and experimental studies document the significance of skin absorption. Parathion should therefore be regarded as a skin exposure hazard.

Diazinon

Skin exposure potential

Diazinon (O,O-diethyl-O-(2-isopropyl-4-methyl-6-pyrimidyl) phosphorothioate) is an organphosphorus pesticide that has mainly insecticidal and acaricidal properties. Thus, it is used mainly as an insecticide in plant growing and veterinary science.

Physicochemical properties

Diazinon is a liquid at room temperature; the boiling point is 83–84°C. It is slightly soluble in water and very soluble in most organic solvents. With its large molecular size (M = 304) and the considerable volatility, this lipophilic compound would mainly penetrate the skin if applied under occlusion.

Experimental data

The dermal LD_{50} is 900 and 455 mg kg^{-1} for male and female rats, respectively, i.e. only about two-fold above values obtained from oral administration (Gaines et al., 1969). Although the mammalian toxicity of pure diazinon is relatively low, that of technical products varies considerably by the batch, probably owing to impurities (Eto, 1979). The toxic effects are due to cholinesterase inhibition, and diazinon is in this respect similar to other commonly used organophosphorus compounds (WHO, 1986).

Human data

Several cases of human diazinon intoxication have occurred (Derache, 1977; Reichert et al., 1977). In the cases recorded in the published literature, dermal absorption was not indicated. However, the absence of evidence in this regard in no way means that percutaneous uptake of diazinon is non-existent or safe. Thus, diazinon is regarded as highly toxic, though less toxic than parathion when absorbed through the skin (ILO, 1983).

Conclusion

Diazinon belongs to the group of organophosphorus pesticides that is known to cause poisoning from skin exposure. Limited experimental evidence confirms the toxicity due to the percutaneous uptake, but definite human data in this regard are lacking. Despite the limited information available, diazinon should be regarded as a skin absorption hazard.

Other compounds

Tetraethyl pyrophosphate (TEPP) has a very low dermal LD_{50} in rats, i.e. 2.4 mg g^{-1}, a value close to those seen with other administration routes (Gaines, 1969). Because of its high toxicity (cholinesterase inhibition), contact with this compound must be avoided.

Sulfotep (tetraethyl dithiopyrophosphate) is a related compound with a higher dermal LD_{50} in the rat, i.e. 65 mg kg^{-1}, five-fold above values seen with oral exposure (Kimmerle and Klimmer, 1974). This compound may occur as a toxic impurity in formulations of diazinon.

With organophosphates having a relatively high degree of water solubility, such as trichlorphon and dimethoate, absorption was faster from liphophilic solvents, such as benzene; less water soluble organophosphates, butonate and phosmet, penetrated the skin more rapidly from acetone or dimethyl sulfoxide (DMSO) solutions (Dedek, 1980).

For phorate (thimet), the dermal LD_{50} in rats was 2.5 mg kg^{-1}, for mevinphos (phosdrin) 4.2 mg kg^{-1} and for ethion 62 mg kg^{-1}; other organophosphates with values less than 100 mg kg^{-1} were schradan, fensulfothion, parathion, and methyl parathion; these values were all similar to those seen with oral dosage (Gaines, 1969). Dimethoate has a dermal LD_{50} value below the oral value, and chlorpyrifos (dursban) has a dermal LD_{50} slightly above (RTECS, 1983). Fenitrothion, merphos and mevinphos were also more toxic after dermal application than oval intake, but small differences in LD_{50} values were seen for many other organophosphates (Gaines, 1969).

Dosing of pesticides on the rear feet of mice caused rapid development of toxicity after exposure to mevinphos (20 min), and the amount needed to cause toxicity was less than with other pesticides; azinphos-methyl (guthion) had the longest latency period of 15 h at a 50-fold higher exposure (Skinner and Kilgore, 1982).

Experimental studies in human volunteers have documented that ^{14}C-labelled azodrin, ethion and azinphos-methyl are absorbed through the skin at a rate similar to parathion and malathion (Feldman and Maibach, 1974).

With methyl demeton, clinical symptoms and cholinesterase inhibition were recorded in workers packaging the pesticide in containers; protective clothing was insufficient, and absorption was considered to occur largely through the skin (Jones, 1982).

Workers entering a cotton field 2 h after application of methyl parathion adsorbed milligram amounts of the insecticides on the skin of hands and arms; decreases in cholinesterase activity confirmed that percutaneous absorption took place (Nemec et al., 1968). Inhalation exposure in workers checking sprayed fields for insect damage is much lower than the dermal exposure (NIOSH, 1976). Methyl parathion, in particular the emulsifiable concentrate, was difficult to remove by laundering of contaminated clothing (Laughlin et al., 1985). More recent studies have confirmed that methyl parathion is absorbed through the skin of spraymen (Kummer and van Sittert, 1986).

Filling and cleaning spraying equipment used for monocrotophos application resulted in decreases in cholinesterase activity and increased excretion of a metabolite, dimethyl phosphate; inhalation exposure was of minimal significance in comparison to skin contact with the pesticide (Kummer and van Sittert, 1986).

Within 72 h after application of chlorpyrifos to the skin of adult rats, the majority of the dose had penetrated the skin, an absorption rate similar to that seen with parathion (Shah et al., 1987).

The solubility characteristics have been calculated for tri-*p*-cresyl phosphate; they approach those of psoriasis scales, thus suggesting the possibility of a consideable dermal uptake (Hansen and Andersen, 1988). However, the related tri-*o*-cresyl phosphate is of particular interest because this compound can cause delayed neuropathy in humans. The permeability constant k_p for application of this compound on human skin *in vitro* is very low, i.e. 0.0025×10^4 cm h^{-1} (Tregear, 1966), thus suggesting that skin penetration will be slow and unlikely to play an important role. However, radioactively-labelled tri-*o*-cresyl phosphate rubbed into the palms of the hands (and left for 3 h) resulted in peak serum levels after 1 h and detectable radioactivity in blood and urine up to 48 h; the total urinary excretion corresponded to 13 and 36% of the amount applied in two subjects (Hodge and Sterner, 1943). Percutaneous absorption was also demonstrated in a dog (Hodge and Sterner, 1943).

References

ABOU-DONIA, M. B. (1981), Organophosphorus ester-induced delayed neurotoxicity, *Annual Review of Pharmacology and Toxicology,* **21,** 511.
ABOU-DONIA, M. B., SIVARAJAH, K. and ABOU-DONIA, S. A. (1983), Disposition, elimination and metabolism of O-ethyl O-4-nitrophenyl phenylphosphonothioate after subchronic dermal application in male cats, *Toxicology,* **26,** 93-111.
ALDRIDGE, W. N., MILES, J. W., MOUNT, D. L. and VERSCHOYLE, R. D. (1979), The toxicological properties of impurities in malathion, *Archives of Toxicology,* **42,** 95-106.
ATALLAH, Y. H., CAHILL, W. P. and WHITACRE, D. M. (1982) Exposure of pesticide applicators and support personnel to O-ethyl O-(4-nitrophenyl) phenylphosphonothioate (EPN), *Archives of Environmental Contamination and Toxicology,* **11,** 219-25.
BAKER, E. L., ZACK, M., MILES, J. W., ALDERMAN, L., WARREN, S. and TEETERS, W. R. (1978), Epidemic malathion poisoning in Pakistan malaria workers, *Lancet,* **i,** 31-33.
BIESBY, J. A. and SIMPSON, G. R. (1975), An unusual presentation of systemic organophosphate poisoning, *Medical Journal of Australia,* **2,** 394-5.
DAS, Y. T., TASKAR, P. K., BROWN, H. D. and CHATTOPADHYAY, S. K. (1983), Exposure of professional pest control operator to dichlorvos (DDVP) and residue on house structures, *Toxicology Letters,* **17,** 95-9.
DEDEK, W. (1980), Solubility factors affecting pesticide penetration through skin and protective clothing. In *Field worker exposure during pesticide application,* eds W. F. Tordoir and E. A. H. Heemstra-Lequin (Studies in environmental science, Vol. 7) (Amsterdam: Elsevier), pp. 47-50.
DERACHE, R. (1977), *Organophosphorus pesticides,* (Oxford: Pergamon).
DIKSHITH, T. S. S., DATTA, K. K. and CHANDRA, P. (1976), 90 day dermal toxicity of DDVP in male rats, *Bulletin of Environmental Contamination and Toxicology,* **15,** 574-9.
DURHAM, W. F., WOLFE, H. R. and ELLIOT, J. W. (1972), Absorption and excretion of parathion by spraymen, *Archives of Environmental Health,* **24,** 381-7.
ETO, M. (1979), *Organophosphorus pesticides: Organic and biological chemistry* (Boca Raton, Fl: CRC Press).
FELDMAN, R. J. and MAIBACH, M. I. (1974), Percutaneous penetration of some pesticides and herbicides in man, *Toxicology and Applied Pharmacology,* **28,** 126-132.
FENSKE, R. A. (1988), Correlation of fluorescent tracer measurements of dermal exposure and urinary metabolite excretion during occupational exposure to malathion, *American Industrial Hygiene Association Journal,* **49,** 438-44.
FISEROVA-BERGEROVA, V. and PIERCE, J. T. (1989), Biological monitoring V. Dermal absorption, *Applied Industrial Hygiene,* **4,** F14-F21.
FRANCIS, B. M., METCALF, R. L. and HANSEN, L. G. (1985), Toxicity of organophosphorus esters to laying hens after oral and dermal administration, *Journal of Environmental Science and Health,* **B20,** 73-95.
FELDMAN, R. J. and MAIBACH, M. I. (1974), Percutaneous penetration of some pesticides and herbicides in man, *Toxicology and Applied Pharmacology,* **28,** 126-132.
FREDERIKSSON, T. (1961), Studies on the percutaneous absorption of parathion and paraoxon-III. Rate of absorption of parathion, *Acta Dermato-Venereologica,* **41,** 353-62.
FREDERIKSSON, T., FARRIOR, W. L. Jr and WITTER, R. F. (1961), Studies of the percutaneous absorption of parathion and paraoxon-I. Hydrolysis and metabolism within the skin, *Acta Dermato-Venereologica,* **41,** 335-43.

GAINES, T. B. (1969), Acute toxicity of pesticides, *Toxicology and Applied Pharmacology*, **14**, 515-34.
HANSEN, C. M. and ANDERSEN, B. H. (1988). The affinities of organic solvents in biological systems, *American Industrial Hygiene Association Journal*, **49**, 301-8.
HAYES, W. J. Jr, MATTSON, A. M., SHORT, J. G. and WITTER, R. F. (1960), Safety of malathion dusting powder for louse control, *Bulletin of the World Health Organization*, **22**, 503-514.
HODGE, H. C. and STERNER, J. H. (1943), The skin absorption of triorthocresyl phosphate as shown by radioactive phosphorus, *Journal of Pharmacology and Experimental Therapeutics*, **79**, 225-34.
HOLMSTEDT, B. (1959), Pharmacology of organophosphorus cholinesterase inhibitors, *Pharmacology Review*, **11**, 567-688.
IARC (1979), *Some halogenated hydrocarbons*, (IARC Monographs on the evaluation of carcinogenic risk of chemicals to man, Vol. 20) (Lyon: International Agency for Research on Cancer).
ILO (1983), *Encyclopaedia of occupational health and safety*, 3rd edn. (Geneva: International Labour Office).
IRPTC (1982), *Parathion*. (Scientific reviews of Soviet literature on toxicity and hazards of chemicals, No. 10) (Moscow: Centre of International Projects, GKNT).
JEGIER, Z. (1974), Health hazards in insecticide spraying of crops, *Archives of Environmental Health*, **8**, 670-674.
JIA-HUAO, S., ZHEN-QIA, W., YI-LAN, W., YI-XIOU, Z., SHOU-ZHEN, X. and XUE-QI, G. (1985), Prevention of acute parathion and demeton poisoning in farmers around Shanghai, *Scandinavian Journal of Work and Environmental Health*, **11**, (Suppl. 4): 49-54.
JONES, K. H., SANDERSON, D. M. and NOAKES, D. N. (1968), Acute toxicity data for pesticides, *World Review of Pesticide Control*, **7**, 135-43.
JONES, R. D. (1982), Organophosphorus poisoning at a chemical packaging company, *British Journal of Industrial Medicine*, **39**, 377-81.
KIMMERLE, G. and KLIMMER, O. R. (1974), Acute and subchronic toxicity of sulfotep, *Archives of Toxicology*, **33**, 1-16.
KNAAK, J. B., YEE, K., ACKERMAN, C. R., ZWEIG, G., FRY, D. M. and WILSON, B. W. (1984), Percutaneous absorption and dermal dose-cholinesterase response studies with parathion and carbaryl in the rat, *Toxicology and Applied Pharmacology*, **76**, 252-63.
KUMMER, R. and VAN SITTERT, N. J. (1986), Field studies on health effects from the application of two organophosphorus insecticide formulations by hand-held ULV to cotton, *Toxicology Letters*, **33**, 7-24.
LAUGHLIN, J., EASLEY, C. and GOLD, R. E. (1985), Methyl parathion residue in contaminated fabrics. In *Dermal exposure related to pesticide use*, (ACS symposium series 273). eds. R. C. Honeycutt, G. Zweig and N. N. Ragsdale, pp. 177-87, (Washington, DC: American Chemical Society).
MAIBACH, H. I., FELDMAN, R. J., MILBY, H. and SERAT, W. F. (1971), Regional variation in percutaneous penetration in man, *Archives of Environmental Health*, **23**, 208-211.
MATHIAS, C. G. T. (1983), Persistent contact dermatitis from the insecticide dichlorvos, *Contact Dermatitis*, **9**, 217-8.
MATSUSHITA, T., AOYAMA, K., YOSHIMI, K., FUJITA, Y. and UEDA, A. (1985), Allergic contact dermatitis from organophosphorus insecticides, *Industrial Health*, **23**, 145-53.
MILBY, T. H. and EPSTEIN, W. L. (1964), Allergic contact sensitivity to malathion, *Archives of Environmental Health*, **9**, 434-437.

NABB, D. P., STEIN, W. J. and HAYES, W. J. Jr (1966), Rate of skin absorption of parathion and paraoxon, *Archives of Environmental Health,* **12,** 501-5.

NEMEC, S. J., ADKISSON, P. L. and DOROUGH, H. W. (1968), Methyl parathion absorbed to the skin and blood cholinesterase levels of persons checking cotton treated with ultra-low-volume sprays, *Journal of Economic Entomology,* **61,** 1740-2.

NIOSH (1976), *Criteria for a recommended standard... Occupational exposure to methyl parathion,* (DHEW (NIOSH) Publ. No. 77-106) (Cincinnati: National Institute for Occupational Health and Safety).

NIOSH (1976), *Criteria for a recommened standard... Occupational exposure to parathion.* (DHEW (NIOSH) publication No. 76-190) (Cincinnati, OH: National Institute for Occupational Safety and Health).

NIOSH/OSHA (1978), Occupational health guideline for EPN. (Cincinnati, OH: National Institute for Occupational Safety and Health).

O'BRIEN, R. D. and DANELLEY, C. E. (1965), Penetration of insecticides through rat skin, *Journal of Agricultural Food Chemistry,* **13,** 245-247.

PETTY, C. S. (1958), Organic phosphate insecticide poisoning, *American Journal of Medicine,* **24,** 467-470.

REICHERT, E. R., YANGER, W. L., RASHAD, M. N., KLEMMER, H. W. and HATTIS, R. P. (1977), Diazinon poisoning in eight members of related households, *Clinical Toxicology,* **11,** 5-11.

ROUDABUSH, R. L., TERHAAR, C. J., FASSETT, D. W. and DZIUBA, S. P. (1965), Comparative acute effects of some chemicals on the skin of rabbits and guinea pigs, *Toxicology and Applied Pharmacology,* **7,** 559-65.

RTECS (1983), *Registry of toxic effects of chemical substances,* (Cincinnati, Ohio: National Institute for Occupational Safety and Health).

SHAH, P. V. and GUTHRIE, F. E. (1977), Dermal absorption, distribution and the fate of six pesticides in the rabbit. In *Pesticide Management and Insecticide Resistance,* eds D. L. Watson and A. W. A. Brown, pp. 547-554, (New York: Academic Press).

SHAH, P. V., MONROE, R. J. and GUTHRIE, F. E. (1981), Comparative rates of dermal penetration of insecticides in mice, *Toxicology and Applied Pharmacology,* **59,** 414-423.

SHAH, P. V., FISHER, H. L., SUMLER, M. R., MONROE, R. J., CHERNOFF, N. and HALL, L. L. (1987), Comparison of the penetration of 14 pesticides through the skin of young and adult rats, *Journal of Toxicology and Environmental Health,* **21,** 353-66.

SHELLENBERGER, T. E. (1980), Organophosphate pesticide inhibition of cholinesterase in laboratory animals and man and effects of oxime reactivators, *Journal of Environmental Science and Health,* **B15,** 795-822.

SKINNER, C. S. and KILGORE, W. W. (1982), Acute dermal toxicities of various organophosphate insecticides in mice, *Journal of Toxicology and Environmental Health,* **9,** 491-7.

TREGEAR, R. T. (1966). *Physical Functions of Skin* (London: Academic).

WESTER, R. C., MAIBACH, H. I., BUCKS, D. A. W. and GUY, R. H. (1983), Malathion percutaneous absorption after repeated administration to man, *Toxicology and Applied Pharmacology,* **68,** 116-119.

WHO/FAO (1978), *Data Sheets on Pesticides No. 2 (Rev. 1). Dichlorvos* (Geneva: World Health Organization).

WHO (1982), *Recommended health-based limits in occupational exposure to pesticides. Report of a WHO Study Group,* (Technical Report Series 67) (Geneva: World Health Organization).

WHO (1986), *Organophosphorus insecticides: A general introduction* (Environmental Health Criteria 63) (Geneva: World Health Organization).

WHO (1989), *Dichlorvos*. (Environmental Health Criteria 79) (Geneva: World Health Organization).

Chapter 13
Organic sulfur compounds

Introduction

The organic sulfur compounds (not considered in other chapters) most frequently considered to be skin absorption hazards are indicated in table 13.1. The most important of these are considered separately below. Carbon disulfide is sometimes considered an inorganic compound, and dimethyl sulfate is an ester of sulfuric acid, but both are considered organic sulfur compounds in this book.

This group of compounds includes some solvents, pesticides and drugs. Some sulfur-containing pesticides are dealt with in other chapters (organophosphorus compounds and carbamates). Probably the most extensive dermal exposure occurs with the pesticides.

An additional phenothiazine compound with a 'skin' denotation in one country is aminazin (chlorpromazine); these drugs can cause contact dermatitis.

Table 13.1. Organic sulfur compounds that are considered a skin hazard in several countries (y = yes, n = no, o = other regulation). Those that are considered in detail are italicized.*

CAS No.	Chemical	Number of countries	FRG	Sweden	USA
Aliphatic sulfur compounds					
75-15-0	*Carbon disulfide*	13	y	y	y
77-78-1	*Dimethyl sulfate*	13	y	o	y
Heterocyclic sulfur compounds					
92-84-2	Phenothiazine	6	n	n	y
115-29-7	Endosulfan	7	o	n	y
78-34-2	Dioxathion	6	o	n	y

*Dimethyl sulfoxide (DMSO) is included in the detailed review below, even if it has only been considered a skin hazard in one country.

Carbon disulfide

Skin exposure potential

Carbon disulfide (CS_2) is produced by reacting charcoal or methane with sulfur vapours. The main use of carbon disulfide is in the production of viscose rayon fibres. It is also used as a soil disinfectant and a fumigant (mainly for stored grain), as a solvent, and in the production of CCl_4 and cellophane. Within the EC, the total production in 1983 was 150 ktonnes, an amount equal to the consumption. Although exposures to this compound are usually mechanically controlled, they may still be considerable (WHO, 1979).

Physicochemical properties

Carbon disulfide is a liquid at room temperature. The boiling point is 46.3°C, and the vapour pressure at 25°C is 360 mm Hg. It is soluble in ethanol, benzene and ether, but only slightly soluble (2.2 g l^{-1} at 22°C) in water. The octanol/water partition coefficient is about 100; log P_{ow} values have been reported as 1.84 and 2.16. The solubility in psoriasis scales is limited (Hansen and Andersen, 1988). A penetration rate of 0.89 mg cm^{-2} h^{-1} was calculated on the basis of physicochemical properties (Fiserova-Bergerova and Pierce, 1989). Absorption through the skin of this lipophilic, small molecule would appear likely, especially when evaporation is inhibited.

Experimental data

When the bodies of rabbits were exposed to 1500 p.p.m. of carbon disulfide vapour while breathing clean air, percutaneous absorption could be measured by levels of carbon disulfide exhaled; these increased for at least 3 h (Cohen et al., 1958). Percutaneous absorption from a 1% emulsion in water was documented in a study where the tails of rats were immersed in carbon disulfide (Izmerov, 1983). Important neurotoxicity, nephrotoxicity and vascular changes have been demonstrated in animal experiments (Beauchamp et al., 1983; WHO, 1981). Studies on pregnant rats also indicate that carbon disulfide may affect the behavioural development and enzyme activities in the offspring (Beauchamp et al., 1983; WHO, 1981).

Human data

Percutaneous absorption has been determined in male volunteers using direct and indirect methods; when a hand was immersed for 1 h in an aqueous carbon disulfide solution of 0.33–1.67 g l^{-1}, a cutaneous absorption rate of 0.23–0.79 mg cm^{-2} h^{-1} was obtained if calculated from the amount of the substance which had disappeared from the solution (Dutkiewicz and Baranowska, 1967). No precautions were described regarding evaporation loss from the solution. When estimated from the amounts exhaled, the percutaneous absorption rate was only about one-tenth of this level (Dutkiewicz and Baranowska, 1967); part of this difference could be due to evaporation from the skin following initial cutaneous uptake. Also of importance is the finding that only 3% of the amount absorbed via the skin is excreted by exhalation, while about 25% of the carbon disulfide absorbed through the lungs is subsequently exhaled (Dutkiewicz and

Baranowska, 1967). Thus, carbon disulfide absorbed through the skin may well be more toxic than a similar amount absorbed through the lungs. Calculations suggested that exposure of a skin area of 360 cm^2 to carbon disulfide would cause an absorption more than 30-fold above inhalation of this compound at the threshold limit value level for the same time period (Fiserova-Bergerova and Pierce, 1989). Further, extensive skin contact may result in blisters which resemble second or third degree burns (Hueper, 1936). As skin penetration is relatively fast, some carbon disulfide absorbed through the lungs may also be eliminated through the skin (Harashima and Masuda, 1962). Industrial experience suggests that poisoning may be due to uptake via the skin (Fairhall, 1957), but most case reports and epidemiological studies do not allow any evaluation of the relative importance of skin absorption. The systemic toxicity mainly relates to effects on the nervous system and the blood vessels (Beauchamp et al., 1983; NIOSH, 1977; WHO, 1981). At chronic, low-level exposures, peripheral neuropathy and increased mortality from arteriosclerosis are the most important adverse effects (Beauchamp et al., 1983; NIOSH, 1977; WHO, 1981).

Conclusions

Carbon disulfide may penetrate the skin, although mainly when evaporation is prevented or when extensive immersion takes place. Available evidence suggests that percutaneous absorption may add to the adverse effects due to occupational carbon disulfide exposure, and it should therefore be regarded as a skin exposure hazard.

Dimethyl sulfate

Skin exposure potential

Dimethyl sulfate is produced by reacting dimethyl ether with sulfur trioxide. The annual production in western Europe in 1983 was estimated at more than 31 000 tonnes. The major use of dimethyl sulfate is as an alkylating agent, mainly for intermediates in the dye, pharmaceutical and perfumery industries.

Physicochemical properties

Dimethyl sulfate is an oily liquid with a boiling point of 188°C and a vapour pressure of less than 1 mm Hg at 25°C. Dimethyl sulfate is soluble in water (28 g l^{-1}), but undergoes hydrolysis; it is miscible with many polar organic solvents and aromatic hydrocarbons, but only sparingly soluble in aliphatic hydrocarbons. Dimethyl sulfate is a strong alkylating agent that reacts with active hydrogen and alkali salts. Thus, dimethyl sulfate would be expected to penetrate the skin, but the rate would be limited by reactions in the skin, thus causing local adverse effects.

Experimental data

Dimethyl sulfate can be absorbed through the skin (WHO, 1985), but no quantitative data are available. Application of 5 ml of dimethyl sulfate to the skin of a rabbit resulted in death after 22 h (Weber, 1902). Immersion of the tails of mice into dimethyl sulfate twice resulted in a 50% mortality (Molodkina et al., 1979). Dimethyl sulfate causes severe effects at the absorption site and systemic toxicity, such as cyanosis, convulsions and coma (WHO, 1985). Sufficient evidence is available on carcinogenicity to animals (IARC, 1987).

Human data

Several reports describe systemic toxicity related to cutaneous exposures to dimethyl sulfate (Balazs, 1934; Littler and McConnell, 1955; Weber, 1902), although inhalation of vapours may have contributed to the toxic effects. Skin contact results in delayed effects (after 1-2 h) which may still develop despite immediate and thorough irrigation and neutralization (Littler and McConnell, 1955; Wang et al., 1988). Systemic effects include hyperthermia, convulsions, delirium, coma and delayed damage to the liver, kidney and myocardium (Roux et al., 1977; WHO, 1985; Wang et al., 1988). Skin contact with the liquid compound results in severe damage; exposure of the skin to dimethyl sulfate vapours may also cause blistering (WHO, 1985). The evidence for carcinogenicity to humans is inadequate, and this compound is considered a class 2A carcinogen (IARC, 1987).

Conclusions

Skin contact with dimethyl sulfate can cause systemic toxicity and severe effects on the skin. Although quantitative data are not available, the serious toxicity and probable carcinogenicity would indicate that dimethyl sulfate should be regarded as a skin absorption hazard.

Dimethyl sulfoxide

Skin exposure potential

Dimethyl sulfoxide $(CH_3)_2S{:}O$, DMSO) is a derivative of lignin and has extraordinary properties as an aprotic solvent, and can bring into solution a variety of inorganic and organic chemicals. Thus, it is used as a solvent for resins, fungicides, dyes and pigments, a reactant for chemical synthesis, an extractant, and a reaction medium to accelerate rates of chemical combination. DMSO also has important uses in dermatological treatment.

Physicochemical properties

DMSO is a liquid at room temperature and has a low vapour pressure. The octanol/water partition coefficient (log P_{ow}) is -1.35, thus indicating the hydrophilic character of the compound. The solubility in psoriasis scales suggests that a considerable dermal uptake could occur (Hansen and Andersen, 1988).

Experimental data

DMSO is absorbed into the skin with great speed (Kligman, 1965). If exposed to a rather concentrated solution of DMSO, the horny layer appears to be totally penetrated within minutes and significant quanitites of DMSO remain in the horny layer for about 2 h; its removal by blood vessels in the subcutaneous tissue seems to be a much slower process than penetration of the horny layer (Kligman, 1965). Studies using experimental animals have documented that the acute toxicity of DMSO is very limited (Willson et al., 1965; Smith et al., 1967; Lazarev, 1976). In several species, at several administration routes, LD_{50} values above 2.5 g kg^{-1} have been obtained. Although DMSO may enhance the penetration of various allergens and thereby promote allergic responses, inhibition may also take place, e.g. in the cases of nickel, chromate and cobalt which are apparently chelated by DMSO. Thus, DMSO generally enhances the penetration of other substances through the skin and thereby also promotes the toxic effects of these substances, but inactivation may occur in certain cases.

Human data

The effects of DMSO are well known from the use of DMSO in topical pharmaceuticals (Scheuplein, 1978). The toxicological significance of DMSO is not due to the systemic effects of the compound *per se* but to its action as a vehicle promoting the penetration of other substances through the skin. This effect is limited to DMSO concentrations below 50% and reaches a maximum at about 90% concentration in water; a decrease in this effect at 100% DMSO is due to the considerable hygroscopicity which induces a water flux in the opposite direction. When the horny layer is almost saturated with DMSO, the skin may be rapidly penetrated by various substances that appear to follow a continuous pathway of the solvent through the skin. Some denaturation seems to take place, but the effect is readily reversible. These cutaneous effects only occur as a result of skin exposure. DMSO can also induce urticaria (Fisher, 1986). Systemic toxicity, such as has been documented in animal experiments at very high dosages, would be unlikely to occur as a result of respiratory exposures in industry, and no exposure limit has been determined.

Conclusion

DMSO has little potential for systemic toxicity, but is a powerful vehicle when applied to the skin, thereby promoting the penetration of other substances and causing augmented toxicity.

Other compounds

Phenothiazine is used both as a drug and a pesticide. Dermal exposures cause photosensitization. Exposure to this compound increases the toxicity of organophosphorus pesticides (Gaines, 1962).

Endosulfan in an aqueous suspension has a dermal LD_{50} of 90 mg kg^{-1} in rabbits; several cases of endosulfan-induced convulsions occurred in production workers, but the significance of skin penetration could not be determined (Ely *et al.*, 1967). Cutaneous exposures may also cause contact dermatitis.

Dioxathion has dermal LD_{50} values of 235 and 63 mg kg^{-1} in male and female rats, respectively (Gaines, 1969). In the rabbit, the dermal LD_{50} is about 100 mg kg^{-1} (Frawley *et al.*, 1963). With LD_{50} values from other administration routes being only slightly lower, this compound appears to be readily absorbed through the skin.

References

BALAZS, T. (1934), Dimethylsulfat-Vergiftung, *Samml Vergiftungsfälle*, **5**, 47–50.

BEAUCHAMP, R.O. Jr, BUS, J.S., POPP, J.A., BOREIKO, C.J. and GOLDBERG, L. (1983), A critical review of the literature on carbon disulfide toxicity, *CRC Critical Review of Toxicology*, **11**, 169–278.

COHEN, A.E., PAULUS, H.J., KEENAN, R.G. and SCHEEL, L.D. (1958), Skin absorption of carbon disulfide vapor in rabbits, *Archives of Industrial Health*, **17**, 164–9.

DUTKIEWICZ, T. and BARANOWSKA, B. (1967), The significance of absorption of carbon disulfide through the skin in the evaluation of exposure. In *Toxicology of carbon disulfide*, eds H. Brieger and J. Teisinger, pp. 50–51. (Amsterdam: Excerpta Medica).

ELY, T.S., MACFARLANE, J.W., GALEN, W.P. and HINE, C.H. (1967), Convulsions in Thiodan workers, *Journal of Occupational Medicine*, **9**, 35–7.

FAIRHALL, L.T. (1957), *Industrial toxicology*, p. 81 (Baltimore: Williams and Wilkins).

FISEROVA-BERGEROVA, V., and PIERCE, J.T. (1989), Biological monitoring, V. Dermal absorption, *Applied Industrial Hygiene*, **4**, F14–F21.

FISHER, A.A. (1986), *Contact dermatitis*, 3rd edn., (Philadelphia, Lea & Febiger).

FRAWLEY, J.P., WEIR, R., TUSING, T. and DUBOIS, K.P. (1963), Toxicologic investigations on Delnav, *Toxicology and Applied Pharmacology*, **5**, 605–24.

GAINES, T.B. (1962), Poisoning by organic phosphorus pesticides potentiated by phenothiazine derivatives, *Science*, **138**, 1260–1.

GAINES, T.B., (1969), Acute toxicity of pesticides, *Toxicology and Applied Pharmacology*, **14**, 515–34.

HANSEN, C.M. and ANDERSEN, B.H. (1988), The affinities of organic solvents in biological systems, *American Industrial Hygiene Association Journal*, **49**, 301–8.

HARASHIMA, S. and MASUDA, Y. (1962), Quantitative determination of absorption and elimination of carbon disulfide through different channels in human body, *Internationales Archiv für Gewerbepathologie und Gewerbehygiene*, **19**, 263–9.

HEUPER, W.C. (1936), Etiologic studies on the formation of skin blisters in viscose workers, *Journal of Industrial Hygiene and Toxicology*, **18**, 432–47.

IARC (1987), *Overall evaluations of carcinogenicity: An updating of IARC Monographs Volumes 1 to 42* (IARC Monographs on the evaluation of carcinogenic risk of chemicals to man (Suppl. 7.) (Lyon: International Agency for Research on Cancer).

IZMEROV, N.F., ed. (1983), Carbon disulfide (IRPTC) Scientific reviews of Soviet literature on toxicity and hazards of chemicals 41). (Moscow: Centre of International Projects, GKNT).
KLIGMAN, A.M. (1965), Topical pharmacology and toxicology of dimethyl sulfoxide — part 1, *Journal of the American Medical Association,* **193**, 140-148.
LAZAREV, N.B. (1976), *Dangerous substances in industry* (in Russian), 6th edn. pp. 396-398, (Leningrad: Chimija).
LITTLER, T.R. and MCCONNELL, R.B. (1955), Dimethyl sulfate poisoning, *British Journal of Industrial Medicine,* **12**, 54-6.
MOLODKINA, N.N., PAVLOVSKAYA, G.S. and DYMOVA, E.G. (1979), Toxicological and hygienic evaluation of the dimethyl sulfate production (in Russian), *Gigrena Truda I Professionalnye Zabolevaniia,* **23**, 28-32.
NIOSH (1977), *Criteria for a recommended standard... Occupational exposure to carbon disulphide* (DHEW (NIOSH) Publication No. 77-156). (Cincinnati, Ohio: National Institute for Occupational Safety and Health).
ROUX, H., GALLET, M., VINCENT, V. and FRANTZ, P. (1977), Poisoning by dimethyl sulfate (clinical and bibliographic study), *Acta Pharmacologica et Toxicologica,* **41**, (Suppl. 2): 428-33.
SCHEUPLEIN, R. (1978), Skin permeation. In *The Physiology and Pathophysiology of the Skin,* ed. A. Jarrett vol. 5, pp. 1693-1730 (London: Academic Press).
SMITH, E.R., HADIDIAN, Z. and MASON, M.M. (1967), The single- and repeated-dose toxicity of dimethyl sulfoxide, *Annals of the New York Academy of Science,* **141**, 96-109.
WANG, Y., XIA, J. and WANG, Q. (1988), Clinical report on 62 cases of acute dimethyl sulfate intoxication, *American Journal of Industrial Medicine,* **13**, 455-62.
WEBER, S. (1902), Über die Giftigkeit des Schwefelsäuredimethylesters (Dimethylsulfates) und einiger verwandter Ester der Fettreihe, *Archiv für Experimentalle Pathologie und Pharmakologie,* **47**, 113-27.
WHO (1979), *Carbon disulfide* (Environmental health criteria 10) (Geneva: World Health Organization).
WHO (1981), *Recommended health-based limits in occupational exposure to selected organic solvents,* (Technical report series 664) (Geneva: World Heath Organization).
WHO (1985), *Dimethyl sulfate* (Environmental health criteria 48) (Geneva: World Health Organization).
WILLSON, J.E., BROWN, D.E. and TIMMENS, E.K. (1965), A toxicologic study of dimethyl sulfoxide, *Toxicology and Applied Pharmacology,* **7**, 104-112.

Chapter 14
Conclusions

This review of mechanisms and evidence for percutaneous absorption indicates that skin penetration will occur for almost all chemicals, at least to some extent. Even with chemicals without a 'skin' denotation in any country, skin contact is not always safe. Likewise, skin contact with a known skin hazard is not always a danger.

The evidence available to a large extent refers to past exposure conditions, and cutaneous exposures to some very toxic chemicals now occur to a limited extent because of restricted use. Also, not all cases of human poisoning are reported in the open literature, and published evidence inevitably only represents the tip of the iceberg. Further, a bias may be present, because adverse effects of industrial chemicals may be more visible if these compounds are also used for topical treatment. For such reasons, 'skin' denotation reflects the past rather than the present, not to speak of the future. To a large extent, the same criticism applies to this review.

In the absence of detailed criteria for prediction of percutaneous absorption hazards, the approach to preventive efforts must be cautious. In view of reported evidence, several chemicals can be regarded as confirmed skin hazards (table 14.1). However, the information also suggests that in some situations, e.g. with dilute solution in certain vehicles, limited skin contact may not represent a hazard.

Thus, with the known hazards, 'skin' denotation should be kept, but the meaning should be: Skin contact must be prevented, unless detailed evaluation shows that limited, unavoidable skin contact will not result in any discernible increase of the total exposure.

For all other chemicals, the following restrictions should apply: Skin contact should be always limited, but extensive skin protection will be necessary only when detailed evaluation shows that skin contact may result in a significant increase of the total exposure.

When evaluating the possible significance of dermal exposures, the following factors should be taken into account:

(1) concentration of the compound and the type of vehicle
(2) extent of skin contact, i.e. the area covered

(3) duration of skin contact and efficiency of removal
(4) regional variability of skin permeability
(5) skin disease and wounds
(6) skin hydration and recent contact with damaging chemicals
(7) occlusion
(8) temperature.

Table 14.1. Industrial chemicals that have caused human toxicity and are known to penetrate the skin in significant amounts.

Acrylamide	Malathion
Acrylonitrile	Methanol
Allyl alcohol	2-Methoxyethanol
Aniline	Methyl bromide
Benzene	Methyl iodide
Carbaryl	Nicotine
Carbon disulfide	p-Nitroaniline
Carbon tetrachloride	Nitroglycerin
Chlorinated naphthalenes	Nitroglycol
Cresols	Paraquat
Cyanides	Parathion
Diazinon	PCBs
Dichlorvos	Pentachlorophenol
N,N-Dimethylacetamide	Phenol
N,N-Dimethylformamide	Phenylhydrazine
1,1-Dimethylhydrazine	Picric acid
Dimethyl sulfate	Tetraethyllead
Dimethyl sulfoxide	Tetryl
Dioxane	Thallium
DNOC	Toluene
Epichlorohydrin	Toxaphene
EPN	2,4,6-Trinitrotoluene
Hydrazine	Xylenes

Index of chemical compounds

Acetone 53
Acetone cyanohydrin 69, 76
Acetonitrile 69
Acridine 91
Acrylamide 6, 25, 69, 73-75, 180
Acrylonitrile 25, 69, 75, 76, 180
Adiponitrile 76
Alachlor 66
Alcohols 16, 21, 47, 52, 53
Aldicarb 145
Aldrin 25, 26, 97, 107
Allyl alcohol 25, 47, 51, 52, 180
Allylamine 69
Allyl 2,3-epoxy-1-propyl ether, see Allyl glycidyl ether
Allyl glycidyl ether 151, 154
Aminazin, see Chlorpromazine
Amitraz 136
2-Aminoethanol, see Ethanolamine
(2-Aminoethyl)-1,2-ethanediamine, see Diethylenetriamine
4-Amino-2-hydroxytoluene 117
Amyl caproate 17
Aniline 11, 25, 113-115, 180
Anisidines 113
Arsenic 38, 44
Arsine 38
Atrazine 146
Azinphos-methyl 157, 166
Azodrin 166
Benzaldehyde 82
Benzene 25, 49, 81-84, 180
1,4-Benzenediamine, see p-Phenylenediamine
Benzene sulfonyl chloride 105
Benzidine 113, 118

Benzonitrile 117
Benzo[a]pyrene 4, 17, 81, 92
Benzoyl peroxide 82
Bianiline, see Benzidine
bis(4-Aminophenyl)methane, see 4,4'-Diaminodiphenylmethane
bis(2-Chloroethyl) ether 25, 59, 65
bis(2-Chloroisopropyl)ether 59
bis(4-Isocyanatocyclohexyl)methane, see Methylene bis(4-cyclohexyl isocyanate)
bis(2-Propyl)amine, see Diisopropylamine
Bromobenzene 98
Bromoform 59, 65
Bromomethane, see Methyl bromide
1,3-Butadiene 48
Butanols 17, 22, 29, 53
2-Butanone, see Methyl ethyl ketone
2-Butanone peroxide, see Methyl ethyl ketone peroxide
2-Butenal, see Crotonaldehyde
Butiphos 158
Butonate 165
2-Butoxyethanol, see Ethylene glycol, monobutyl ether
1-(2-Butyoxyethoxy)-2-ethanol, see Diethylene glycol, monobutyl ether
2-Butoxyethel acetate, see Ethylene glycol, monobutyl ether acetate
1-Butylamine 69, 76
Butylamines 25, 69
2-(n-Butyl)aminoethanol 69
Butyl carbitol, see Diethylene glycol, monobutyl ether
Butyl cellosolve, see Ethylene glycol, monobutyl ether
t-Butyl chromate 37

2-sec-Butyl-4,6-dinitrophenol, see Dinoseb
Butylene oxide, see Tetrahydrofuran
t-Butyl hydroperoxide   48
t-Butyl peracetate   48
2-sec-Butylphenol   82
4-t-Butylphenol   82
Cadmium   38
Calcium cyanamide   38
Calcium cyanide   39
Camphechlor, see Toxaphene
Captafol   135, 145
Captan   145
Carbaryl   7, 25, 135–138, 180
Carbitol, see Diethylene glycol, monethyl ether
Carbofuran   136, 145
Carbon disulfide   25, 171–173, 180
Carbon tetrachloride   25, 59–61, 65, 180
Cellosolve, see Ethylene glycol, monoethyl ether
Chlordane   25, 26, 97, 106
Chlordecone   105
Chlorinated biphenyls, see PCBs
Chlorinated camphene, see Toxaphene
Chlorinated dibenzofurans and dioxins   105
Chlorinated naphthalenes   22, 25, 97, 100, 101, 180
Chlorinated paraffins   66
Chloroanilines   114
Chlorobenzene   104
o-Chlorobenzylidene malononitrile   114
2-Chloro-1,3-butadiene, see Chloroprene
Chlorodiphenyl oxides   98
3-Chloro-1,2-epoxypropane, see Epichlorohydrin
2-Chloroethanol   25, 59, 64
Chloroethene, see Vinyl chloride
Chloroethylene, see Vinyl chloride
Chloroform   20, 60, 64, 65
4-Chloro-2-methylaniline, see 4-Chloro-o-toluidine
2-Chloro-1-methylbenzene, see o-Chlorotoluene
Chloronaphthalenes   97
4-Chloronitrobenzene, see p-Nitrochlorobenzene
p-Chlorophenol   97

2-(2-Chlorophenylmethylidene) propanedinitrile, see o-Chlorobenzylidene malononitrile
Chloroprene   25, 26, 59, 65
Chlorpromazine   171
Chlorpyrifos   158, 165, 166
o-Chlorotoluene   97
4-chloro-o-toluidine   114
Chromium   38
Coal tar   81, 91
Cobalt   37
Copper   37
Creosote   91
Cresols   17, 25, 81, 89–91, 180
Crotonaldehyde   48
Cumene   81, 91
Cyanides   37, 180
Cyanogen   38
Cyclohexane   49, 91
Cyclohexanone   81
Cyclohexanone peroxide   82
Cyclohexylamine   113
Cyclonite   121
Cyclotrimethylenetrinitramine, see Cyclonite
Cyhexatin   38
2,4-D   105
DDT   97, 104
DDVP, see Dichlorvos
Decaborane   37, 44
Demeton   25, 26, 157
Demeton-methyl   25, 157
2,4-Diaminoanisole   117
4,4'-Diaminodiphenylmethane   114, 118
Diazinon   25, 157, 164, 165, 180
Dibenzoyl peroxide, see Benzoyl peroxide
Diborane   38
1,2-Dibromo-3-chloropropane   60
1,2-Dibromoethane, see Ethylene dibromide
Dibromomethane, see Methylene bromide
Dibutyl ether   48
Di-t-butyl peroxide   48
Dibutyltin   38
3,4-Dichloroaniline   114
1,2-Dichlorobenzene   97
3,3'-Dichlorobenzidine   114, 118
2,2'-Dichlorodiethyl ether, see bis(2-Chloroethyl) ether

Dichlorodiphenyltrichloroethane, see DDT
1,1-Dichloroethane 60
1,2-Dichloroethane, see Ethylene dichloride
2,2-Dichloroethenyl dimethylphosphate, see Dichlorvos
alpha-Dichlorohydrin 60
beta-Dichlorohydrin 60
Dichloromethane, see Methylene chloride
2,4-Dichlorophenol 16
2,4-Dichlorophenoxyacetic acid, see 2,4-D
1,3-Dichloro-2-propanol, see *alpha*-Dichlorohydrin
2,3-Dichloro-1-propanol, see *beta*-Dichlorohydrin
1,3-Dichloropropene 60
Dichlorvos 25, 157–159, 180
Dicrotophos 158
Dicyclohexyl peroxide 82
Dieldrin 25, 97, 107
Diethanolnitrosamine, see *N*-Nitrosodiethanolamine
Diethylamine 69
N,N-Diethyl-2-aminoethanol 25, 69, 76
Diethylene glycol, monobutyl ether 55
Diethylene glycol, monoethyl ether 55
Diethylene glycol, monomethyl ether 54, 55
Diethylenetriamine 69, 77
1,4-Dihydroxybenzene, see Hydroquinone
2,4-Diisocyanatotoluene, see Toluene-2,4-diisocyanate
Diisopropylamine 69, 76
Dilauroyl peroxide 48
Dimethoate 165
3,3′-Dimethoxybenzidine 118
N,N-Dimethylacetamide 20, 25, 26, 69, 71–73, 180
Dimethylanilines, see Xylidines
N,N-Dimethylaniline 25, 113, 117
7,12-Dimethylbenz[*a*]anthracene 92
Dimethylbenzenes, see Xylenes
o-Dimethylbenzidine, see *o*-Tolidine
alpha,alpha-Dimethylbenzyl hydroperoxide 82
O,O-Dimethyl *O*-(2,2-dichlorovinyl) phosphate, see Dichlorvos

N,N-Dimethylformamide 20, 25, 26, 28, 29, 69–71, 180
1,1-Dimethylhydrazine 25, 135, 142, 143, 180
1,2-Dimethylhydrazine 136
Dimethylnitrosamine 69, 77
Dimethyl phthalate 81
Di(2-methyl-2-propyl) peroxide), see Di-*t*-butyl peroxide
Dimethyl sulfate 25, 171, 173, 174, 180
Dimethyl sulfoxide, see DMSO
1,3-Dinitrobenzene 25, 121, 130
4,6-Dinitro-*o*-cresol, see DNOC
2,4-Dinitrophenol 121
Dinitrotoluene 25, 121, 131
Dinoseb 131
1,4-Dioxane 25, 151, 153, 154, 180
Dioxathion 171, 176
4-Diphenylamine 113
Diphenyl ether, see Phenyl ether
Dipropylene glycol, monomethyl ether 47
Disulfoton 158
DMSO (dimethyl sulfoxide) 11, 20, 24–26, 171, 174, 175, 180
DNOC (4,6-dinitro-*o*-cresol) 25, 121, 129, 130, 180
Dursban, see Chlorpyrifos
EGDN (1,2-ethanedioldinitrate, ethylene glycol dinitrate), see Nitroglycol
Endosulfan 171, 176
Endrin 25, 97, 107
Epichlorohydrin 25, 151–153, 180
EPN (*O*-ethyl-*O*-(4-nitrophenyl)phenylbenzene thiophosphonate) 25, 157, 161, 162, 180
1,2-Ethanediol, see Ethylene glycol
1,2-Ethanediol dinitrate, see Nitroglycol
Ethanol 16–18, 20, 53
Ethanolamine 69
Ethion 157, 165
2-Ethoxybutanol, see Ethylene glycol, monobutyl ether
2-Ethoxyethanol, see Ethylene glycol, monoethyl ether
1-(2-Ethoxyethoxy)-2-ethanol, see Diethylene glycol, monoethyl ether
2-Ethoxyethyl acetate, see Ethylene glycol, monoethyl ether, acetate

2-Ethoxypropanol, see Ethylene glycol, monopropyl ether, acetate
2-Ethoxypropylacetate, see Ethylene glycol, monopropyl ether, acetate
Ethyl acetate 48
Ethyl acetoacetate 48
Ethyl acrylate 25, 47
Ethylamine 69
Ethylbenzene 26, 81, 91
Ethylene chlorohydrin, see 2-Chloroethanol
Ethylenediamine 69
Ethylene dibromide 11, 25, 26, 40, 59, 66
Ethylene dichloride 40, 59, 60, 64, 65
Ethylene glycol 48, 53
Ethylene glycol dinitrate, see Nitroglycol
Ethylene glycol, monobutyl ether 25, 47, 55
Ethylene glycol, monobutyl ether, acetate 48, 54
Ethylene glycol, monoethyl ether 25, 47, 55, 56
Ethylene glycol, monoethyl ether, acetate 25, 47, 55, 56
Ethylene glycol, monomethyl ether 25, 47, 48, 54–56
Ethylene glycol, monomethyl ether, acetate 25, 47, 55
Ethylene glycol, monopropyl ether 55
Ethylene glycol, monopropyl ether, acetate 55
Ethyleneimine 25, 135, 145
Ethylene oxide 151
Ethylene sulfide 151
Ethylmercury 38
N-Ethylmorpholine 135, 146
Fenamiphos 158
Fenitrothion 158, 165
Fensulfothion 165
Fenthion 158
Ferbam 136
Folpet 145
Fonofos 158
Formic acid 48
Furfural 24–26, 151, 154
Furfuryl alcohol 151, 154
Glycol ethers 48, 54–56
Guthion, see Azinphos-methyl

Halogenated hydrocarbons 21
HC Blue No. 1 117
Heptachlor 25, 97, 106, 107
n-Heptane 49
Hexachloro-1,3-butadiene 60
gamma-Hexachlorocyclohexane, see Lindane
Hexachlorocyclopentadiene 98
Hexachloroethane 25, 59, 65
Hexafluoroacetone 60
Hexafluoropropanone, see Hexafluoroacetone
Hexamethylphosphoric acid triamide 158
n-Hexane 26, 47, 49
2-Hexanone, see Methyl n-butyl ketone
Hydrazine 25, 135, 141, 142, 180
Hydrogen cyanide 25, 37, 39, 40
Hydrogen sulfide 38
Hydroquinone 92
Hydroxybenzene, see Phenol
2-Hydroxyethyl acrylate 48
2-Hydroxymethylfuran, see Furfuryl alcohol
Iodomethane, see Methyl iodide
Isobutyronitrile 76
Isocyanates 21
Isocyanatomethane, see Methyl isocyanate
Isophorone di-isocyanate 135, 145
Isopropyl alcohol 48
N-Isopropylaniline 113
Isopropylbenzene 91
Isopropyl palmitate 16, 17
Kepone, see Chlordecone
Ketones 53
Lead 38, 44
Lindane 25, 97, 106
Malathion 5, 7, 25, 157, 159–161, 163, 166, 180
Malononitrile 69
Manganese 38
p-Menthene 91
Mercury 25, 26, 37, 38, 73
Merphos 165
Methacrylonitrile 69
Methanol 16, 20, 25, 47, 49–51, 180
Methomyl 135, 145
Methoxyanilines, see Anisidines

2-Methoxyethanol, see Ethylene glycol, monomethyl ether
1-(2-Methoxyethoxy)-2-ethanol, see Diethylene glycol, monomethyl ether
2-Methoxyethyl acetate, see Ethylene glycol, monomethyl ether, acetate
Methoxyethylmercury 38
1-(2-Methoxypropoxy)-2-propanol, see Dipropylene glycol, monomethyl ether8
Methyl acetate 48
Methyl acrylate 25, 47
Methylamine 69
Methyl amyl alcohol, see Methyl isobutyl carbinol
2-Methylaniline, see o-Toluidine
N-Methylaniline 25, 113, 117
Methyl bromide 25, 59, 61–63, 180
Methyl n-butyl ketone 48, 53
Methyl cellosolve, see Ethylene glycol, monomethyl ether
Methylchloroform, see 1,1,1,-Trichloroethane
2-Methylcyclohexanone 25, 81, 91
Methyl demeton 166
Methylene bromide 65
Methylene chloride 59, 60, 64, 65
4,4′-Methylene-bis(2-chloroaniline) 113, 118
Methylene-bis(4-cyclohexyl isocyanate) 136
Methylenedianiline, see 4,4′-Diaminodiphenylmethane
Methylene dibromide, see Methylene bromide
Methylene dichloride, see Methylene chloride
Methyl ethyl ketone 26, 53
Methyl ethyl ketone peroxide 48
Methyl glycol, see Ethylene glycol, monomethyl ether
Methylhydrazine 135
Methyl iodide 25, 59, 63–64, 180
Methyl isobutyl carbinol 25, 47, 53
Methyl isobutyl ketone 48
Methyl isocyanate 24, 25, 135, 145
Methylmercury 38, 43
Methyl methacrylate 28, 48
N-Methylmorpholine 136
Methyl parathion 25, 28, 157, 165, 166

2-Methylpentane 49
4-Methyl-2-pentanol, see Methyl isobutyl carbinol
4-Methyl-2-pentanone, see Methyl isobutyl ketone
Methylphenols, see Cresols
(2-Methyl-2-propyl) peracetate, see t-Butyl peracetate
Methylstyrene 81
N-Methyl-N,2,4,6-tetranitroaniline, see Tetryl
Mevinphos 25, 157, 165
MOCA, see 4,4′-Methylene-bis(2-chloroaniline)
Monocrotophos 166
Morpholine 25, 135, 146
Myrcene 91
Naphthalene 81, 91
1-Naphthylamine 113
2-Naphthylamine 113
1-Naphthyl-N-methylcarbamate, see Carbaryl
Nicotine 25, 135, 138, 139, 180
Nitramine, see Tetryl
2-Nitro-4-aminophenol 117
p-Nitroaniline 25, 113, 116, 117, 180
Nitrobenzene 11, 25, 121, 125, 126
4-Nitrobiphenyl 121
p-Nitrochlorobenzene 25, 121, 130
Nitroglycerin 25, 28, 121, 123, 124, 180
Nitroglycol 7, 25, 122, 123, 180
p-Nitrophenol 16
2-Nitro-p-phenylenediamine 117
N-Nitrosodiethanolamine 17
N-Nitrosodimethylamine, see Dimethylnitrosamine
Nitrotoluene 22, 25, 121, 130
n-Nonane 49
n-Octane 49
Octanol 11, 16
Organophosphorus pesticides 13, 21, 165
2-Oxetanone, see beta-Propiolactone
Paraquat 25, 135, 139, 140, 180
Parathion 25, 157, 162–166, 180
PCBs 22, 25, 97–100, 180
Pentaborane 38
Pentachlorophenol 25, 97, 101–103, 180
Peracetic acid 48

Perchloroethylene   59, 60, 64, 65
Permethrin   106
Phenol   16, 17, 25, 81, 88, 89, 180
Phenyl ether   82
Phenothiazine   171, 176
Phenylamine, see Aniline
p-Phenylenediamine   24, 25, 113, 117
Phenylhydrazine   25, 135, 143, 144, 180
Phorate   157, 165
Phosdrin, see Mevinphos
Phosmet   165
o-Phthalodinitrile   117
Picric acid   25, 121, 128, 129, 180
Polychlorinated biphenyls, see PCBs
1,2-Propanediol dinitrate, see Propylene glycol 1,2-dinitrate
1,2,3-Propanetriol trinitrate, see Nitroglycerin
1-Propanol, see Propyl alcohol
2-Propanol, see Isopropyl alcohol
Propargyl alcohol   24, 25, 47, 53
2-Propeneamide, see Acrylamide
2-Propenenitrile, see Acrylonitrile
2-Propen-1-ol, see Allyl alcohol
2-Propenylamine, see Allylamine
2-Propenylbenzene, see Methylstyrene
(2-Propenyloxy)methyloxirane), see Allyl glycidyl ether
beta-Propiolactone   151
Propyl alcohol   16, 48
N-(2-Propyl)aniline, see N-Isopropylaniline
2-Propylbenzene, see Cumene
Propylene glycol 1,2-dinitrate   121
Propyleneimine   135, 145
2-Propyn-1-ol, see Propargyl alcohol
Pyrene   91
Pyrethrum   82
Pyridine   135
Pyrrolidine   135
Quinoline   135
Schradan   158, 165
Sodium cyanate   38
Sodium cyanide   39
Sodium fluoroacetate   25, 37, 44
Silver   37
Styrene   26, 81, 91
Sulfotep   157, 165
Systox, see Demeton

2,4,5-T   98, 105
TCDD, see 2,3,7,8-Tetrachlorodibenzo-p-dioxin
TEDT, see Sulfotep
TEPP   157, 165
1,1,2,2-Tetrabromoethane   11, 59
2,3,7,8-Tetrachlorodibenzo-p-dioxin   105
2,3,7,8-Tetrachlorodibenzofuran   105
1,1,2,2-Tetrachloroethane   25, 59, 64, 65
Tetrachloroethene, see Perchloroethylene
Tetrachloroethylene, see Perchloroethylene
Tetrachloromethane, see Carbon tetrachloride
Tetrachlorophenols   97, 105
Tetraethyl dithiopyrophosphate, see Sulfotep
Tetraethyllead   25, 37, 38, 40–43, 180
Tetraethyl pyrophosphate, see TEPP
2,2,3,3,-Tetrafluoro-1-propanol   60
Tetrahydrofuran   151
Tetramethyllead   25, 37, 38, 40
Tetramethylsuccinonitrile   69, 76
Tetryl   25, 121, 124, 125, 180
Thallium   25, 37, 42–43, 180
Thimet, see Phorate
Tin   38
TNT, see 2,4,6-Trinitrotoluene
o-Tolidine   114
Toluene   15, 25, 29, 81, 84–86, 180
p-Toluenediamine   118
Toluene-2,4-diisocyanate   145
o-Toluidine   25, 26, 113, 117
Toxaphene   25, 97, 103, 104, 180
Tribromomethane, see Bromoform
1,1,1-Trichloroethane   29, 60, 64, 65
1,1,2-Trichloroethane   25, 59, 64, 65
Trichloroethene, see Trichloroethylene
Trichloroethylene   26, 60, 64, 65
Trichloromethane, see Chloroform
Trichlorophenols   16, 97, 105
2,4,5-Trichlorophenoxyacetic acid, see 2,4,5-T
Tri-o-cresyl phosphate   158, 166
Tri-p-cresyl phosphate   166
Triethylamine   69
alpha,alpha,alpha-Trifluorotoluene   97
Trinitrophenylmethylnitramine, see Tetryl
2,4,6-Trinitrophenol, see Picric acid

2,4,6-Trinitrotoluene 25, 121, 127, 128, 180
Triphenyltin 38
UDMH, see 1,1-Dimethylhydrazine
Vinylbenzene, see Styrene
Vinyl chloride 60
Warfarin 151
Xylenes 15, 21, 24, 25, 29, 81, 86–88, 180
Xylidines 25, 113, 117
Zinc 37